PROBLEMS
OF
PLACE-NAME STUDY

PROBLEMS
OF
PLACE-NAME STUDY

Being a Course of Three Lectures delivered at King's College under the auspices of the University of London

BY

A. MAWER

CAMBRIDGE

At the University Press

1929

CAMBRIDGE
UNIVERSITY PRESS

University Printing House, Cambridge CB2 8BS, United Kingdom

Cambridge University Press is part of the University of Cambridge.

It furthers the University's mission by disseminating knowledge in the pursuit of education, learning and research at the highest international levels of excellence.

www.cambridge.org
Information on this title: www.cambridge.org/9781107455634

© Cambridge University Press 1929

This publication is in copyright. Subject to statutory exception and to the provisions of relevant collective licensing agreements, no reproduction of any part may take place without the written permission of Cambridge University Press.

First published 1929
First paperback edition 2014

A catalogue record for this publication is available from the British Library

ISBN 978-1-107-45563-4 Paperback

Cambridge University Press has no responsibility for the persistence or accuracy of URLs for external or third-party internet websites referred to in this publication, and does not guarantee that any content on such websites is, or will remain, accurate or appropriate.

CONTENTS

PREFACE

The Lectures here printed are given, except for a few minor alterations, exactly in the form in which they were delivered at King's College under the auspices of the Board of English Studies in the University of London. Many of the matters discussed in them might well have been treated at greater length and in fuller detail. Time did not allow of such treatment during the lectures themselves, and to attempt to expand them now or even to make them coldly impersonal would inevitably destroy such unity of thought and feeling as may have been attained under the circumstances of their original composition. The lectures as delivered were not interrupted by reference to sources or authorities, and such references have been reduced to a minimum in the printed text. For all statements with regard to places in Buckinghamshire, Bedfordshire, Huntingdonshire, Worcestershire, the North Riding of Yorkshire, and Sussex, the reader is referred to the volumes on those counties published by the English Place-name Society. There and in the Introductory volumes of that Survey (all fully indexed) will be found the material upon which all observations not only upon the place-names of these counties but also upon scattered place-names throughout the country are based. The notes to these lectures confine themselves strictly to matter not referred to in those volumes.

A. M.

August, 1929

INTRODUCTION

When the University of London did me the honour of inviting me to lecture on "Some aspect of Place-name Study," I hesitated for some little time about the wisdom of accepting the invitation, and that for more than one reason. We are only at the beginning of the great enterprise of a historical survey of the place-names of England. Only a few corners of the vast field have as yet been thoroughly explored, and it is dangerous to come to any general conclusions on so slender a basis. If one draws premature conclusions, there is always the natural danger that new facts may soon come to light which will effectively overthrow them, but there is also a more subtle danger. One's own views on delicate and difficult problems, which ought at this stage still to be in a fluid condition, may tend to become prematurely hardened and make one less sensitive to new evidence when it comes to light. There was also the further difficulty that in the absorbing work of the Survey during the last six or seven years, almost all one's discoveries had been made for the Survey and had already found their natural place in the volumes published by it. Any lectures one could give would inevitably include a good deal of repetition of things already to be found in them.

On the other hand, one realised that in those volumes, which include whole masses of material, early and late, and information of every kind, historical, linguistic and topographical, there was serious danger that the public at large and even the specialist in this or that particular field of historical or linguistic study, might not be able to see the wood for the trees, and so it was felt it might serve a useful purpose to sort and re-arrange some of this material, some of this information, in ways which would show, at least in one or two fields of interest, just what was being discovered. A good many of the results of the Survey, *e.g.* those embodied in the second of these lectures, are already established beyond controversy, except in points of minor detail. On other matters, where judgment must still be left in suspense, it may be helpful to set forth the evidence as so far known, and to indicate some of the considerations which have to be borne in mind in settling the problems which lie before us.

In the end hopes outweighed fears, the invitation was accepted and the lectures were written, but as one wrote them, one realised more than ever how little was entirely one's own and how much one had profited during the last few years by the work of others. The Survey is a co-operative effort. Almost every page of its volumes bears witness to the unselfish help given by scholars both lay and professional. For many of its best discoveries it has been indebted to scholars like Professor Ekwall, Dr Ritter and Professor Zachrisson. The debt to them, so far as is humanly possible, is set forth in the volumes themselves. My debt to my co-editor, Professor Stenton, has had of necessity to remain unexpressed in those volumes, for every page of which we are jointly responsible, but now that I am presenting some of the results of our joint work in my own name, I can only say that without his help at every stage the Survey would never have been undertaken and no one of its volumes could have been written. In these lectures I am still further indebted to him for some of the examples of early ME personal names quoted in the third lecture.

ABBREVIATIONS

ASC	*Anglo-Saxon Chronicle.*
BCS	Birch, *Cartularium Saxonicum*, 3 vols., 1885–93.
Du	Dutch.
EHR	*English Historical Review.*
EPNS	English Place-name Society Publications.
IPN	*Introduction to the Survey of English Place-names*, 1923.
KCD	Kemble, *Codex Diplomaticus Aevi Saxonici*, 6 vols., 1839–48.
LGer	Low German.
LOE	Late Old English.
MDu	Middle Dutch.
ME	Middle English.
MHG	Middle High German.
MLGer	Middle Low German.
ModHG	Modern High German.
ModLG	Modern Low German.
NFr	Norman French.
Norw	Norwegian.
ODan	Old Danish.
OE	Old English.
OED	*Oxford English Dictionary.*
OFr	Old French.
OFris	Old Frisian.
OHG	Old High German.
OIr	Old Irish.
ON	Old Norse.
OS	Old Saxon.
OSw	Old Swedish.
PN	Place-names.
ZONF	*Zeitschrift für Ortsnamenforschung.* (In progress.)

LECTURE I

Racial Settlement

Of all the varied problems and possibilities which the study of place-names offers there are none perhaps which command wider attention than those concerned with early racial settlements in this island. The historical texts for that period are scanty and their authority has often been impugned. Within the last half-century or so we have realised however that there are still two sources of information open to us, as yet largely unworked, which should be invaluable in supplementing and checking the information found in the historical texts. These sources are archaeological remains and place-names. The importance of archaeology is illustrated by the entirely new theories as to the line of approach of the West Saxons—by the estuary of the Wash and the Icknield Way—which have been put forward by Mr Thurlow Leeds on the basis of archaeological evidence. Place-name students have not as yet advanced any theory so bold as this on the basis of their research. If they ever do, the important thing for all to remember—historian, archaeologist and place-name student alike—is that any reconstruction which is now advanced must be consistent with the evidence derived from all these varied sources.

The first and perhaps the most popular problem is that of the fate of the Britons at the hands of their English conquerors. When Professor Ekwall published his *Place-names of Lancashire* in 1922, one realised what place-names might tell us on this score from his survey of the distribution in that county of names of British origin or containing British elements. He showed how in that county those names tended to form groups situated in areas which are or had once been wild hill-country, marshland, or forest-land, suggesting clearly the conditions under which a British-speaking population might have survived for some time after the Anglian settlement.

Intensive study of other counties has yielded nothing parallel to this. In Buckinghamshire the Celtic element is confined to *Brill*, in which the first element is the same Celtic word for "hill" which we have in *Bredon* in Worcestershire and *Breedon* in Leicestershire; *Chetwode*, an example of a common hybrid type with British *cet* followed by English *wood*; *Bernwood* Farm in Claydon, the last relic of the name of an extensive forest, containing the British equivalent of Welsh *bryn*, "hill"; with the more doubtful *Datchet* in the extreme south-east of the county and *Panshill*, near Brill, which may contain the same wood-name as *Pancett* Wood in Wiltshire. A lost wood-name *Moreȝyf*, *Moreyf*, a parallel to the Staffordshire *Morfe*, is found in

Westbury in the 13th century, while *Brayfield* in this county and *Brafield*-on-the-Green in Northamptonshire, ten miles away, on the other side of Yardley Chase, may contain the old Celtic name for that area as their first element. Hill and woodland names alone therefore, apart from the usual river-names, show evidence of Celtic origin.

Bedfordshire yielded nothing certainly Celtic either in place- or field-names, beyond one or two additional stream-names, notably a second Severn, which long survived in *Severne Ditche* in Bedford, another *Camel* river on which stands *Campton*, and a tiny *Humber* river. In Huntingdonshire yet another *Humber* was found. Apart from that we have one hill-name, *Lattenbury* in Godmanchester, OE *Lodona beorg*, which may contain the old British name for the fen country which lies at its foot.

One might have expected a rich harvest of Celtic names in Worcestershire as we approach the Welsh border. What do we find? Three Celtic hill-names—*Bredon*, *Carton*, and *Malvern*; five names containing British *cruc*, "hill, barrow"; *Pensax* containing Welsh *pen*, "headland"; a trace of Welsh *mynydd*, "hill," in a lost *Minton* in Eastham-on-Teme; a Celtic marsh-name in *Corse*, Celtic *Dorne*, probably denoting a fort on the Fosse Way; many river-names which, after the fashion of the west and south-west, have at times given name to villages on their banks (*e.g. Kyre*, *Laughern*); *Worcester* itself; and a tiny residuum of names like *Mamble* and *Tardebigge*, as yet unsolved, which are probably Celtic.

That is the total among some 1700 names which were studied.

Conditions in the North Riding of Yorkshire are more difficult to assess, for we do not know how far Celtic names may have survived the Anglian invasion but ultimately given place to new Scandinavian ones. The number of ultimate Celtic survivals is exceedingly small. Notable are *Catterick* and *Crayke*, but apart from these we only have river-names and a few names like *Dinnand*, of uncertain etymology, which may really be Celtic. If we may judge by the analogy of the neighbouring county of Durham, where Scandinavian influence was never very strong, it is not likely that Celtic names were much more numerous in this area even in Anglian days.

Last and clearest in its evidence is the county of Sussex. Here, beyond a stray river or two, we have no certain Celtic place-names at all. It looks as if the statement of the Chronicle with reference to the storming of *Andredesceaster* or Pevensey, that "the South Saxons slew all who dwelled therein and there was not one Briton who survived," may have been only too true of a great deal of the conquest of Sussex.

Taking the place-name evidence as a whole, it is clear that, in these counties at least, we can build little or nothing upon it in support of the idea of an extensive survival of a British population, still less of a British-speaking population, after the Saxon and Anglian Conquest.

That view may be correct, but it must be supported, if at all, on other grounds. It should perhaps be added that this view of the preponderatingly English character of our place-nomenclature is strengthened in many ways by the great survey of English river-names recently completed by Ekwall, in which he demonstrates English origin for a much larger proportion even of these names than had hitherto been allowed or suspected. The evidence of all other counties, so far as it has been gathered for the Survey, tends in the same direction. Even in Devon, for which the material is almost complete, the Celtic element is, in proportion to the whole, surprisingly small.

Before we pass on to the English settlements themselves, something should perhaps be said about the group of names—*Walton, Walcot(t), Walworth* and the like—which have usually figured somewhat prominently in any discussion of the relations of the English with the conquered Britons. The generally accepted interpretation of these names has been that they denote "farms, cottages, enclosures of the *Welsh* (OE gen. pl. *Weala*) or Britons." Place-name students have however long since recognised that two important qualifications of this general view were needed. The first is that, as the term *wealh* came ultimately to be used in English of a serf generally, not necessarily of British birth, still less of British speech, we must not press these names unduly as evidence for actual survival of a British population. So far as they are really old they probably do suggest such a survival, but we have no means of gauging when these names arose. Some may date from a time when slaves were almost entirely of English birth. The other qualification is that noted (very tentatively) in the treatment of Walton in Buckinghamshire and Walton in Huntingdonshire, viz. that if the early ME forms show no evidence for a medial *e* between the *l* and the *t*, *i.e.* if we have *Waltone* rather than *Waletone*, then one cannot interpret these names as going back to OE *wēala-tūn* but must rather interpret them as containing a first element *weall*, "wall," or *weald*, "forest."

Zachrisson, in his stimulating paper, *Romans, Kelts and Saxons in Britain*, recently published, has made a more serious attack on the use of these names as evidence for Celtic survival. In the case of the *Walton*-names at least, he is inclined to think that practically none of them go back to OE *Wealatun*, and he would interpret them, and many of the other *Wal*-names besides, as all containing either *weall* or *weald*. ME forms in *Waletone* he would take to be bad spellings, probably Anglo-French, for *Waltone*. The only positive evidence of any strength adduced for this view is the case of a *Wealtun* in an original document of the 10th century, which Zachrisson identifies with a place called *Waletune* in the Suffolk Domesday. This identification is however exceedingly doubtful. The OE form is in the will of Aelfhelm Polga, who bequeathes property situated in Cambridgeshire, West Suffolk and

North Essex. It is most unlikely that his property included one manor right away in the east of Suffolk, for the Domesday *Waletune* is Walton near Felixstowe. *Wealtun* is much more likely to be some lost place in the west of the county[1]. At any rate it is clearly unsafe to build any large superstructure upon this one piece of evidence. The theory of an Anglo-Norman spelling seems equally doubtful. There is no evidence for such a development, and a widespread one too, of an *e* between *l* and *t*. We have plenty of examples of *sealt* and *wealt* (from earlier *weald*) in place-names. Anglo-French spellings with *saut* and *waut* are numerous; none have been found with *salet* or *walet*. One must still believe, I think, that forms with medial *e* are of real significance in the history of these names, while fully admitting that Zachrisson has done excellent service in making us scan all these names much more closely and carefully than we have perhaps always done in the past.

A word may be added as to the alternative possibilities of *weald* and *weall* which are, as we have seen, fully admissible and indeed probable in those names for which we have persistent ME *Waltone* or, as in the case of the *Wealtun* just discussed, have the luck to have an unambiguous Old English form. *Weall* is clearly right wherever we can interpret *Walton* as "farm by the wall,"

[1] This point has been made independently by Professor Ekwall in a paper published in *Studia Neophilologica* i, 106–7.

e.g. Walton Savage (DB *Waltone*) in Shropshire might, if it were rather nearer Wall Town (DB *Walle*), where there is a Roman encampment, contain OE *weall* as suggested by Zachrisson (*op. cit.* 74), but it is a good two miles away up hill and down dale, and one must in any case beware of assuming too readily the existence of such walls in England now or at any time. The interpretation of *wealltun* as "farm or enclosure with or made by a wall" is of course a possible one (cf. the common *Stanton*), but we cannot, as has been suggested, take *weall* in the sense "steep hill, cliff," for this is a purely poetic development, not found in prose.

When we turn to the story of the English invaders themselves the counties with which we have to deal can be grouped as follows according to historical tradition: (i) Bedfordshire, Huntingdonshire and the North Riding of Yorkshire, as Anglian settlements, the two former in the district later known as Mercia, the last in Northumbria; (ii) Sussex, as an independent settlement by the South Saxons; (iii) Worcestershire, as a settlement made in the first instance by a people called the *Hwicce*, who were closely related to if not identical with the West Saxons, but later passed under the dominion of Mercia; (iv) Buckinghamshire, a settlement clearly occupying a border position between West Saxons, Anglians, and Middle Saxons. What has the study of the place-names of these counties done to reinforce, correct, or make more precise these general views?

In Worcestershire the existence of *Wichenford*, six miles north-west of Worcester, probably bears witness to the presence of the mysterious *Hwicce*. The place was perhaps so called because it was the first ford in *Hwiccan* territory reached by a traveller from the territory of the *Magasætan* (the present Herefordshire) to the west. *Pensax* must contain the name of the West Saxons themselves, and the most likely interpretation of the name is that put forward by Ritter (and independently by Zachrisson), viz. that it means the "Pen-Saxons," *i.e.* the Saxons who settled on this particular *pen* or headland. They would be appropriately so called as against their neighbours to the west of the Teme who dwelled in what has always been regarded as Anglian territory.

Worcestershire passed under Mercian rule in the 7th century. How far this meant an Anglian immigration into the district is a moot point. One or two place-names suggest that some such immigration did take place. In Worcestershire we have three examples—*Newbold*-on-Stour, *Wychbold* and *Boughton* near Worcester—of the word *bold*, "building", only found elsewhere as a place-name element in what is known to have been Anglian territory. It is used in the laws of Ine in the compound *boldgetal*, "collection of houses", but as the element is never found in place-names in what is known beyond question to have been West Saxon territory, it was presumably already an archaic term among the West Saxon settlers, and was never used in their place-name formations. *Boughton* is a compound of *bold* and *tun*, apparently the Mercian equivalent of the common Northumbrian *Bolton* from *boðltun*, both alike meaning "enclosure with a building on it."

More interesting than these are *Phepson* in Himbleton, six miles north-west of Worcester, and *Whitsun* Brook in Flyford Flavell, three or four miles from the eastern border of the county. Phepson is OE *Fepsetnatun*, *i.e.* farm of the *Fep*-settlers, whatever be the interpretation of *Fep-*, and must certainly be brought into relation with the *Feppingas* of Bede, who lived in Middle Anglia, and with the *Færpingas* of the Tribal Hidage, who probably belonged to Western Northamptonshire. A migration westwards of some of these *Feppingas* is the most reasonable explanation of the name Phepson, *Fepsetnatun* being just the name which we should expect to arise under such circumstances, the suffix *sæte* being added to the first element of a name, regardless of its significance. Even more clearly, Whitsun Brook (OE *Wixena broc*) must be associated with the *East-* and *West-Wixena* of the Tribal Hidage, who were settled in the neighbourhood of the Lincolnshire fens. Some small band of them must have migrated to Worcestershire.

The story of the settlement of Buckinghamshire has been considerably affected by the new alignment given to the West Saxon advance in 571 by the discovery that *Lygeanburg* is *Limbury* near Luton and not *Lenborough* near Buckinghamshire. This gives a line Limbury-Aylesbury-

Bensington, which cuts through the county in such a way as to suggest the conquest of South and Central Buckinghamshire alone. In support of this distinction between West Saxon Buckinghamshire on the one hand and Anglian Buckinghamshire on the other, approached from the valley of the Ouse, we have a good deal of place-name evidence. The distinctively Anglian *botl*, "building," is found only in *Bottle* or *Botolph* Claydon, well to the north of the line indicated, and in the lost *Newbottle* in the extreme north-west of the county. Further, for what it is worth (*v. infra* 20 ff.), we may note the Anglian *c* of *Calverton* in the north as against the West Saxon *ch* of *Chalvey* and *Chawley* in the south of the county, both containing the common word *c(e)alf*, "calf." On the other hand, in the south we have one example of OE *ȳfre*, "slope," in *Iver*, and four of *ōra*, "bank," elements which have not as yet been found in distinctively Anglian territory, except possibly one or two examples of *ora* in Herefordshire.

The counties of Bedfordshire, Huntingdonshire, and the North Riding of Yorkshire show no such complexity of racial settlement, at least at the stage of the English conquest, and their English nomenclature offers fewer points of interest. The nomenclature of the North Riding however offers one feature which is worthy of comment. In this county we have numerous examples of place-names (some twenty-five in all) formed from the dative plural of a common noun, governed originally by the preposition *at*.

Among them we may note the following: three examples of *Angram* from OE *anger*, "grass-land"; *Downholme* from OE *dūn*, "hill"; two cases of *Leatham* from OE *hlið*, "slope"; *Healam* from OE *healh*, "nook, corner"; *Acomb* from OE *āc*, "oak"; *Lealholme* from OE *lǣla*, "twig"; *Yarm* from OE *gear*, "yair, fishing enclosure"; *Beadlam* from OE *boðl*, "building"; *Skiplam* from OE *scipen*, "cow-byre"; and the common *Coatham* and *Wykeham* from OE *cot*, "cottage," and *wīc*, "dwelling, dairy farm." There are several examples of *Newsham* and one or two other similar formations from the dative plural of *hūs*, "house." These may be either Scandinavian or Anglian, but as *Newsham* is also found in Northumberland and Durham in places where Scandinavian influence is unlikely, Anglian origin is very probable. In addition to these we have a good number of dative plural formations of definite Scandinavian origin, such for example as *Airyholme* and *Eryholme* from ON *erg*, "shieling." Some at least of these may have taken the place of earlier Anglian names of similar formation.

The distribution of these dative plural formations in other English counties is of interest: *cotum*, in the various forms *Co(t)ton*, *Cotten*, *Co(a)tham*, *Cottam*, is widespread, and is found in Durham, the North and East Ridings of Yorkshire, Lancashire, Cheshire, Shropshire, Staffordshire, Derbyshire, Nottinghamshire, Lincolnshire, Leicestershire, Warwickshire, Oxfordshire, Northamptonshire, Huntingdonshire and Cambridgeshire; *wicum*, in the various forms *Wykeham*, *Wykin*, *Wicken*, makes a bad second in popularity and is found in Leicestershire, Warwickshire, Northamptonshire and possibly in Cambridgeshire. Apart from these we get only entirely sporadic occurrences of such formations. "At the willows" is found as *Willen* in North Buckinghamshire, as *Welwyn* and *Willian* in Hertfordshire, "at the streams" (OE *flēot*, "stream") in *Flitton* in Bedfordshire. *Herne* in the same county is identical with *Harome* in the North Riding and goes back to a lost OE **hār*, denoting perhaps a "stony place." All these are in the South Midlands. It is only when we reach the North Midlands, and still more when we come to Northumbria, that these formations become really common. Fifteen have been noted in the West Riding, ten in Lancashire, seventeen in Northumberland and Durham[1], quite apart from names of Scandinavian origin, and all showing the same wide variety of type as was illustrated above under the North Riding. Staffordshire, Derbyshire, Nottinghamshire and Lincolnshire show an intermediate stage in the frequency of this formation; four examples have been noted in Staffordshire, six in Derbyshire, six in Nottinghamshire and seven in Lincolnshire.

[1] In some of these cases the present writer, in his *Place-names of Northumberland and Durham*, hesitated between explaining them as examples of OE *hām*, weakened in ME forms to *om*, *um* or of OE dat. plurals in *-um*. Further examination of the evidence suggests that the latter is the right explanation of almost all of these names and possibly of all.

Taking the distribution as a whole, it is noteworthy that these formations are confined to Anglian England. The only exceptions that have been noted in the whole of the rest of England are the common use in Devon of *Hayne* from the dative plural of OE *(ge)hæg*, "enclosure," with a few other dative formations in the same county; the difficult *Carhampton* in Somersetshire, which probably goes back to the *æt Carrum* of the Anglo-Saxon Chronicle; a *Wicken* and a doubtful *Ashen* in Essex, where the terminal *en* makes its appearance at a somewhat late date. Within Anglia itself it is clear that this method of place-name formation was much more popular among the Northumbrians than among the Mercians, but the gradual shading off of this feature as we travel south suggests that this was a matter of local fashion rather than of any racial difference among the settlers themselves. It is perhaps worth noting that this place-name type does not seem to be found among the Anglians of Norfolk and Suffolk.

The last, and in some ways the most interesting county with which we have to deal is Sussex. The two pieces of historical tradition in the matter of its settlement are as follows. First we have the precise statements of the Anglo-Saxon Chronicle with reference to its invasion by Aella and his sons Cymen, Wlencing and Cissa in 477, culminating in the sack of Pevensey by Aella and Cissa in 491, an invasion which was, according to the tradition, entirely independent of the invasion of Wessex on the one hand and of Kent on the other. Secondly, we have the inferences that may be drawn from the phrase used in the ancient annals preserved by Simeon of Durham in his *Historia Regum, s.a.* 771, where he speaks of the subjugation of the *gens Hastingorum* by Offa of Mercia. By itself this might be taken merely as a loose alternative phrase for Sussex as a whole, but when we note that even as late as 1011 the Anglo-Saxon Chronicle (E-text) can, in a list of counties ravaged by the Danes, speak of Kent and Sussex and *Hæstingas* and Surrey, it is clear that the terms are not alternative. It looks as if in Eastern Sussex there may have been a settlement entirely distinct from that made by those whom, for want of a better term, we may call South Saxons. How far is this assumption borne out by place-name evidence?

We may note in the first place that the name of the *Hæstingas* themselves is preserved not only in the present *Hastings*, earlier *Hæstingaport* or *Hæstingaceaster*, but also in *Hastingford*, near Hadlow Down, in the rape of Pevensey, and in *Hastingleigh* in South-east Kent, not far from the Sussex border. What the area settled by them was we do not know, but if difference of place-nomenclature is of any significance in this matter then it would seem that the ultimate scope of influence of the *Hæstingas*, though not necessarily their exact area of original settlement, was the three eastern rapes. Certain place-name elements are used in those rapes which are not found in the three western ones. The clearest and most important examples of these are as follows: OE **hāþ*, "heathland," as distinct from OE *hǣþ*, surviving in *Hoathly, Hodore* and sixteen other names for which we have early record, as well as in numerous names known only in their modern form; OE *glind*, "fence, enclosure," found in *Glynde* and in six other names; OE *sceorf*, "steep slope," found in *Hodshrove* and *Hoadsherf*; OE *stiorf*, a woodland term of uncertain meaning found four times; OE *bing*, "hollow," found twice; OE **e(t)sce* meaning possibly "pasture-land," found seven times. Finally there is the curious *gill*, "deep narrow valley," curiously reminiscent of the North Country *ghyll*, which goes back to OE **gyll* and is widespread throughout the eastern Wealden area[1]. Of these *glind* and *e(t)sce* are not known elsewhere, the latter being confined to the extreme south-east of the county, while the others are found also in the Wealden area in Kent.

Distinctive western features are much rarer. The most important are four or five examples of *Fyning* or *Vining*, an obscure woodland term, probably a derivative of OE *fīn*, "wood-heap"; two examples of an OE *sængel*, probably denoting "brushwood," found in *Singleton* and in the old name *Sengle* for part of Cowdray Park; and two examples of an OE **rispe* or **ripse* denoting "shrubs, briars[1]". More important is the use in *Nyewoods*, in two examples of *Nyetimber*, and in the early forms of *Newtimber*, just

[1] For further notice of some of these words, *v. infra* 16 ff.

over the eastern border of Bramber rape, of an OE form ***nīge*, "new," instead of the usual *nīwe*. This is a distinction which belongs to the pre-English period and is not a dialectal difference arising in English itself; cf. OFris *ny*, *ni*, OS *niwi* and *nigi*, OHG *niuwi*. It is noteworthy that this form seems to be found in *Newtimber* in Warblington in South-east Hampshire, where we have fluctuation between *Ny-* and *New-*, and in *Niton* in the Isle of Wight and *Newchurch* in that island, with the same fluctuation in the early forms. On the other hand, the *in* or "home" tithing of Henfield, just over the western border of Lewes rape, is called *Intendyng* in 1374, preserving the distinctive form *tende* for *tenth*, only found elsewhere in Kentish, a distinction of form again going back to pre-English sound-developments; cf. OFris *tianda*, *tienda* (Du *tiende*), OS *tehande*.

In the use of common place-name elements it is perhaps worth noting that *hamtun*, unknown in Kent, is found only in the two westernmost rapes, while in the rape of Hastings we have a large preponderance of *ham(m)*-names over *tun*-names, such as is without parallel in the rest of the county.

All this evidence tends materially to strengthen the case for believing in a two-fold settlement of the county, but the distinction between East and West Sussex must not be pressed too hard or made too definite. There are a good many terms peculiar to Sussex (with an extension into Kent or Surrey in some instances) which cannot thus be divided between east and west. A few such are OE **stumbel*, "stump," known elsewhere only in Surrey and Kent (three examples); *scydd*, "shed," or the like, a Wealden term found also in Kent; *smēagel*, "burrow" (three examples); the mysterious *Sunt(e)*, found once in Shipley and once in Lindfield and in the name Sompting; *spic*, another word of uncertain meaning, found twice in Sussex and known also in Kent[1]. There is also the curious topographical use of OE *strǣl*. The only recorded sense in Old English is "arrow," but we get three places called *Streel(e)* in the county with well-marked topographical characteristics. One is in a long arrow-like projection of Pulborough parish along the banks of the Arun, the second is on a narrow ridge and the third in a narrow valley. This would suggest that in Old South Saxon the word had developed the sense "long narrow strip of land."

There are also certain general peculiarities to which one might call attention but which, owing to the uncertainty as to their age, it would be unsafe to use in arguing questions of primitive settlement. Two noteworthy cases are the numerous *et*-formations from tree and plant-names, such as *Birchett, Haslett, Rushout*, and the common use of such terms as *Eastout, Northin, Northup*. The former are occasionally found elsewhere though to nothing like the same extent. Examples of the latter have not so far been observed elsewhere. To these may perhaps be added four

[1] For further notice of some of these words, *v. infra* 64 ff.

examples of *Homewood, i.e.* apparently wood near the manor-house, a term only noted elsewhere in the Surrey *Holmwood.* All these go back to the 13th century at least, one of them to the 12th, but we do not know how much older they may be.

Finally, on this question of a distinctive South Saxon place-name vocabulary, we may note the survival of three terms reflecting the social conditions of the early settlement. We have two examples of OE *gebūrscipe,* denoting some association or community of *būras* or peasants, a term known only from the laws of Edward the Elder, and one example of *seten,* denoting apparently "land taken into cultivation," known only from the laws of Ine, but surviving in *Seaton* in Boughton Aluph in Kent. More striking are the four examples, scattered over the county, of OE *morgen-giefu,* the *morgen-gabe* or morning-gift, given by old Teutonic custom by the bridegroom to the bride on the day of the marriage. This term has only been met with once elsewhere, viz. in Surrey, and suggests that in these parts the morning-gift often took the form of a piece of land.

Place-name evidence tends then to confirm what may be inferred from historical tradition, viz. that Sussex stands somewhat apart from the rest of England in the story of its settlement. It shares some of its peculiarities with Surrey and even more with Kent, but it should be borne in mind that this community of usage, so far as it is confined to the Wealden area, and it is there that it is most common, may be due to an extension of an original South Saxon term into the neighbouring Kent or Surrey districts of the Weald, or of Kent and Surrey terms into Sussex, at a comparatively late date. A clear illustration of this possibility is afforded by the extraordinary frequency of the element *fold* in the rape of Arundel and, to a smaller degree, in the rape of Bramber, *i.e.* just in those rapes which are in direct contact with Surrey, the county where *fold*-names are specially common. More definite conclusions will not be possible until full analyses have been made of the place-name material of all the four counties which border on Sussex.

One general problem remains for consideration before we leave the question of the English settlement. In recent years, following the lead of Ekwall in his *Contributions to the History of Old English Dialects*, much use has been made of place-name material for the study of the historical phonology of English dialects. It is well to realise that, important as these studies are in the history of our language, they can only throw light to a very limited extent upon questions of racial settlement. The criteria used consist to a large extent of sound-changes which arose within the Old English period itself, many of them developing comparatively late in that period, though some at least may be the outcome of latent differences going back to a much earlier time. At times the distribution of these phenomena agrees with the racial distribution suggested by historical tradition and confirmed as we have just seen by the evidence of place-name usage and vocabulary, but again and again it cuts right across it. Ekwall, in the essay just mentioned, shows how Anglian England stands in clear contrast to Saxon and Kentish England in its treatment of *a* before *l* followed by a consonant, so that in the former we get forms like *Wold* and *Calverton* (from *calf*) against *Weald* and *Chalvey* (from *cealf*) in the latter, but the position with regard to Anglian *wælle*, "spring," as against Saxon and Kentish *wielle, wylle, welle,* is by no means so clear. Passing from west to east we have ME forms *wille, wulle, welle* which tend to bind Wessex, Sussex, Surrey, Middlesex, Kent and Essex together, but Anglia falls into two halves, Western Mercia which shows *walle* (cf. the modern *Cresswall, Heswall*) and Eastern Mercia and East Anglia which show *welle.* This last form may, as has been suggested, go back to OE *wælle,* with a different sound-development, and represent differences of racial settlement between Eastern and Western Mercia, but such differences do not seem very likely. More striking is the case of OE *y*. If we follow its ME development we must bind together Sussex, Kent, Essex, Suffolk as *e*-counties; Surrey, Wessex, the West and Central Midlands as *u*-counties; the North and the East Midlands as *i*-areas. Or again, in the general treatment of mutated *a* before *n*, yielding ME *e* as in *fenne* in some counties and *a* as in *fanne* (*e.g. Bulphan* in Essex) in others, the greater part of Sussex must go with Kent, Essex, Hertfordshire and Bedfordshire against the rest of England.

All this suggests that the ultimate distribution of phonological features as we are able to examine them in their Middle English manifestations is of little or no value as furnishing criteria for settling questions of early racial settlement. We do not know sufficient about the date and circumstances under which such distribution arose. It is unfortunate that in this matter such Old English charters as have survived give us little or no help. The vast majority (whether preserved in their original form or as transcribed in cartularies it makes little difference) have come down to us in a form of Anglo-Saxon of markedly West Saxon character, and even the lists of boundary-points are written out, not in the local dialect, but in a language which seldom differs seriously from the West Saxon *koine*. A good example of this is furnished by the forms of *Wansford* in Huntingdonshire. All the post-Conquest forms and the present-day pronunciation show that the first element in this name is the Anglian *wælm*, "spring, stream." The one Anglo-Saxon form that we have, found in a 10th-century charter preserved in the Peterborough cartulary, is the West Saxon *wylm*. In similar fashion Domesday and early Middle English evidence generally make it clear that Sussex was a county in which the dialectal development of *y* to *e* was widespread if not universal. Only three *e*-forms have been noted in any Anglo-Saxon charters for the county, namely *Bexlea* and *brecge* for West Saxon *byxe*, "box-tree," and *brycg* in the Bexhill charter (BCS 208) and *heðe* for *hyðe*, "harbour," in the newly identified Eastbourne charter (BCS 1124). In the Bexhill charter there are several other words which do not show this change, and it is not found at all in the Ambersham charter (BCS 1114) or the two versions of the Washington charter (BCS 834, 1125), both of which are preserved in much better texts. It has been suggested that BCS 702, a grant of land at *Derantune*, is concerned with Durrington in Sussex[1]. This is by no means certain, but even if it is correct and *Derantun* thus shows OE *e* for *y*, there are some half-dozen other examples of *y* in the charter which have kept the West Saxon form.

From this digression let us turn to the story of the Scandinavian settlements in England as reflected in the place-name material. Here, so far as the counties as yet surveyed are concerned, we have in the first place a clear illustration of the widely different ways in which the Viking invasions affected different parts of the country, which were all grouped later under the common term *Danelaw*. At the one extreme, in the character and extent of that influence, stands Buckinghamshire, at the other the North Riding of Yorkshire, with Bedfordshire and Huntingdonshire occupying an intermediate position, but approximating, as one might expect, much more to the

[1] Karlström, *OE Compound PN in -ing*, 147 n. 2.

conditions prevailing in Buckinghamshire than to those of the North Riding.

Surveying the nomenclature of Buckinghamshire in general fashion one might very much doubt the statement found in a document of the time of Henry I that Buckinghamshire was one of the fifteen Danelaw counties. It was Ekwall who first called attention to *Skirmett,* the name of a hamlet in Hambleden, with its tell-tale initial *sk*, as a possible example of Scandinavian influence. When in the course of the Survey early forms of this name were found, it became clear that the name went back to a ME *skirmote,* the earliest form being *la Skirmote* in 1307. This seemed beyond question to be a Scandinavianising of the OE *scīr(ge)mōt,* "shire-" or "district-moot," with substitution of Scandinavian *sk* for English *sc*, pronounced as *sh.* Now Skirmett is half a mile from *Fingest,* which in its old form is clearly *þing-hyrst, i.e.* wood or wooded hill associated with a *þing.* But *þing* is the common Anglo-Scandinavian term for a political assembly, and it becomes at once clear that the two names point to the same phenomenon, viz. the holding in this neighbourhood of a county- or hundred-moot which was not only sufficiently Scandinavian in character to be called by the somewhat common term *þing,* but also, and this is much more remarkable, to the presence of persons speaking some Scandinavian or Anglo-Scandinavian dialect who so impressed their influence as to change the English *scir* to the Anglo-Scandinavian *skir* in a way for which we can only find a parallel if we go as far north as the wapentake of *Skyrack* in Yorkshire, where OE *scīr-āc*, "shire-oak," has undergone a similar transformation. It should be added that these names receive further significance when we note that *Turville*, the next parish to Fingest on the farther side from Hambleden, contains as its first element the Anglo-Danish personal name *þyre*, doubtless that of a Dane who attended, possibly presided over, the shire-moot in question. Apart from these names the evidence from place-names for the presence of Scandinavian settlers in Buckinghamshire is almost negligible. *Ravenstone* and *Turweston* in the extreme north-west of the county, on the Northamptonshire border, contain Anglo-Scandinavian personal names, while *Owlswick* in Princes Risborough possibly contains the Scandinavian personal name *Úlfr*. Apart from these, some half-dozen Scandinavian personal names have been noted in field-names, chiefly in the northern part of the county, while in connexion with *Brand's Fee* in Hughenden we have a curiously persistent use in a Buckinghamshire family of Scandinavian personal names down to the 13th century. The Scandinavian *holm* is fairly common in field-names, but it is found only in the north of the county and the word is so common in ME that it proves nothing.

Buckinghamshire as a whole is a good illustration of a Danelaw county where there was no settlement *en masse* or general dividing up of the land. Danish influence on its nomenclature is practically confined to the official and land-owning classes and Danish settlers were probably very few in number.

One would expect something different in turning to Bedfordshire. There is much to be found in the Chronicle in the 9th, 10th, and 11th centuries telling of the activities of the Danes in this county, and we hear of all the chief *jarls* (A-text) or *holds* (D-text) who obeyed Bedford, *i.e.* presumably the Danish leaders who bore sway over the district later known as Bedfordshire, making submission to Edward the Elder in 918. Here however, just as in Buckinghamshire, the mark left by the Danes upon the nomenclature of the county was very slight, and the most interesting piece is official in character. The hundred of Manshead, occupying the south-west corner of the county, includes the parish of *Tingrith* and the actual meeting place of the hundred was near some fields still called *Manshead,* close to the boundary brook which separates the parishes of Tingrith and Eversholt. The brook from which *Tingrith* takes its name is clearly a *thing-rithe, i.e.* one by which a *þing,* such as we described under Fingest (*supra* 24), was held. Doubtless *þing* was the regular term in Bedfordshire for a hundred-assembly during the period of Danish domination.

Apart from this, the evidence for Scandinavian settlement is slight. *Renhold* and *Stagsden,* in the north of the county, may contain Scandinavian personal names *Hráni* and *Stakkr*, *Francroft* in Sharnbrook in the north-west corner, and *Clipstone* in Manshead hundred, from personal names *Fráni* and *Klyppr*, certainly do so. In medieval field and minor names we have *bigging* in Tempsford, a fair number of *holms*, a *toft* in Stotfold in the east of the county, a *toft* and a *wang* in Sharnbrook which fit in with the *Francroft* mentioned above, one *wro* (from ON *vrá*, "corner") in Milton Bryant, the next parish to Tingrith, and one *beck* in Clophill in the centre of the county. All one can say is that if the names are distributed very thinly they are also distributed very evenly. There is certainly no distinction to be drawn between Bedfordshire east and Bedfordshire west of the Alfred-Guthrum line which went up the Lea, across to Bedford, and then up the Ouse. One's whole impression must be that there cannot at any time have been any extensive or deep-rooted settlement of Danes in the county.

In passing into Huntingdonshire we advance still farther into the Danelaw. At the outset we have the important fact that two of the four Huntingdonshire hundreds bear Scandinavian names, and one at least can only have been given by Scandinavian-speaking people. The southernmost hundred of *Toseland* takes its name from the parish of that name. *Toseland* contains as its second element ON *lundr*, "heathen grove," or, to use the phrase of Reginald of Durham in interpreting the *lund* of Plumbland in Cumberland, *nemus paci donatum*. The first part is the Scandinavian personal name *Toli*, the bearer of which may have been the very earl *Toli* or *Toglos* from this district who fell at the battle of Tempsford in 922. This *Toli*, whoever he was, clearly still practised his old heathen religion.

The other hundred is that of *Normancross*. The early forms show this to be *Norðmannes-cros*, *i.e.* the cross associated with a man called the *Norðman* or "Norseman," for such was the clear use of this term in England. Such a name would be given to some Viking settler who, in contrast to the general body of Danish settlers in these parts, had come from Norway. The word *cross* is itself of course also of Scandinavian origin, an Irish word picked up by the Vikings during their raids and settlements in Ireland. *Normancross* is doubtless a much younger name than *Toseland*.

Danish personal names in Huntingdonshire place-names are disappointingly few. *Keyston* in the extreme west of the county derives from one *Ketill*, and *Clack* Barn in the next parish from one *Klakkr*, and there is a sprinkling of Scandinavian personal names in field and minor names generally. Among the major place-names we have *holm* in *Bromholme* and *Port Holme* near Huntingdon in the north, on the edge of the fens; two examples of *þorp*, *Upthorpe* in Spaldwick and Ellington *Thorpe* in the west of the county, and another *lund* in *Holland* (earlier *Haulund*) in the wooded area of the Riptons. *Coppingford* in the heart of the county is *Copemaneforde* in Domesday. The first element is the late OE *coupmanna* (gen. pl.) from ON *kaupmanna*, "merchants". This ford was clearly used by traders who, either because they were themselves Danish or because they traded with a Danish population, came to be described by this Anglo-Scandinavian term. Among minor names we have a *beck* in Eynesbury, many *holmes*, several *tofts* and *wangs*, and one *lathe*, from ON *hlaða*, "barn." The sum-total is not large, but one feels that a definite change has taken place. The word *lund* was never adopted into English speech, and names containing it can only have arisen among a Scandinavian-speaking population. *hlaða* was adopted, but only in the North and the North Midlands, and its presence in Huntingdonshire points in the same direction, while the two *thorpes* are just of the type one expects Danish thorps to be, viz. small villages due to colonisation from a larger one.

Historical tradition would lead us to look for something quite different when we turn to Yorkshire. For Bedfordshire and Huntingdonshire we have no statement comparable with that of the Chronicle, telling us (*s.a.* 876) that "Healfdene portioned out the lands of the Northumbrians and they (*i.e.* the Danes) tilled it and made a living by it." Here we evidently have a settlement *en masse*, a taking over of whole districts from their English occupants. Doubtless the occupation was not equally intensive in all the districts alike. One slight illustration of this is perhaps to be found in the use of the name *Ingleby*. This can only mean "village" or "farm of the English," and it is a name which would have no significance except in areas where the population was preponderatingly Scandinavian, with occasional survival of groups of the earlier inhabitants. The name is unknown in the West and East Ridings, but three examples are found in the wapentake of Langbargh West in the North Riding, which, taking everything into account, is probably the most intensely Scandinavian of the three Ridings.

The term *riding* itself is of course of Scandinavian origin and so is the *wapentake*, which here takes the place of the *hundred*, but curiously enough only one among the early wapentakes bears a distinctively Scandinavian name, that of *Halikeld*. The assembly of the wapentake was naturally known as a *þing*, hence we get the first example of a *þing-vǫllr*, the "meeting-place of the *þing*," in a lost *Thingwall* not far from Whitby, and also in *Fingay* Hill in West Harlsey, a round-topped hill standing out prominently from the level land of the parish, earlier *þing-haugr*, "assembly hill." Close by the latter is *Landmoth* in Leake, from OE *land-(ge)mot*, "land-meeting," doubtless another hill where the wapentake or riding might hold its court, the pair furnishing an interesting parallel to the *Fingest* and *Skirmett* discussed above.

In Yorkshire we are faced for the first time with the interesting problem of the relative part played by Danes and Norsemen in the Viking settlements. How important that distinction might be can be realised when we remember the internecine feuds in the 10th century between Danish and Norse princes for the control of the kingdom which centred round York, or the fact that the poem found in the Chronicle, *s.a.* 942, celebrating the deliverance of the Five Boroughs, is not one of praise for the freeing of the Boroughs from Danish domination but for the freeing of those Danish boroughs from heathen Norwegian tyranny[1]. It is this twofold invasion by Norse and Danish Vikings which gives significance to the four *Normanbys* and the three *Danbys* found in the North Riding. Such names can only have been given in districts where there was sufficient admixture of Danes and Norsemen to make it either necessary or worth while to distinguish between settlements made by one or the other. These names have no parallels in the East Riding, except for the isolated *Danthorpe*. It is true that in the West Riding we have one *Normanton* and a *Denaby* and three or four examples of *Denby*. These are found however in the south part of the Riding, where we have little reason to suspect the presence of any large body of Norse invaders. The very form of the names however probably gives the clue to the matter. The *Denby*-names show a first element *Dena*, the OE genitive plural of *Dene* in contrast to the ON *Dana*, the genitive plural of *Danir*, while *Denaby* (DB *Denegebi*) comes ap-

[1] See on this point *The Redemption of the Five Boroughs* in EHR 38, 551–557.

parently from the OE genitive plural *Denigea*, a yet more distinctively English form. Normanton and the Denbys and Denaby were doubtless so called in contrast to the English population around them in an area which was never Scandinavianised to anything like the same extent as the North and East Ridings or the north-western part of the West Riding.

For distinguishing the part played by Danes and Norsemen in the Viking settlement of Yorkshire we now have more than one test at our disposal. Björkman and others have taught us long since to distinguish between Danish *thorpe, clint, hulme, brink* on the one hand and Norse *gill, scale, slack, breck* on the other, to mention but a few of these distinctions. Our knowledge of East Scandinavian as against West Scandinavian personal nomenclature is as yet all too scanty, but we are now able to distinguish at least a few Scandinavian place-names as of definitely East Scandinavian origin from the presence in them of personal names which, so far as our knowledge goes, are of distinctively East Scandinavian character, *e.g. Tofi* as found in *Towthorpe* or *Klakkr* as found in *Claxton*, but Ekwall's work on *Scandinavians and Celts in North-western England* opened up a new field and showed that the Norse settlements, owing to the close associations of the Norsemen with Ireland and the Isle of Man, show a strong Irish colouring in their place and personal nomenclature. To this there is no parallel in those settlements which we know to have been of purely Danish origin. Whenever then in the North Riding we get traces of such an inversion compound as *Hillbraith* for the more usual *Braith(s)-hill*, or place-names compounded with an Irish personal name, such as *Melsonby* from OIr *Maelsuithan*, or place-names containing the Irish-Scandinavian *erg*, "shieling," such as *Eryholme* (dat. plural), or a district in Domesday which contains tenants bearing purely Irish names, we may suspect the presence of Norse settlers.

Following up these various clues, Smith has been able to show quite clearly that while the wapentakes of Bulmer, Birdforth, Ryedale and Pickering, forming a half-circle around York, are preponderatingly Danish, Gilling East and West, Richmondshire, Langbargh East and West and, rather unexpectedly, Whitby Strand show very definitely the marks of extensive Norse settlements. Further, the distribution definitely suggests that the Norse settlers came in from the west, swarming over from Lancashire, Cumberland, and Westmorland, rather than from the east. That in spite of this they did make their way far to the east is seen by the occasional traces of their presence in areas that are for the most part clearly Danish. There is an *Airyholme* (from *erg*) in Hovingham in Ryedale, an *Irton*, *i.e.* farm, of a Norseman or Norsemen nicknamed the *Irishman* (or *-men*) in Seamer, not far from the North Sea, and there was an *Irton*, now lost, near Coxwold, while the *erg*-names go over the border into the East Riding in *Argam*, *Arram* and *Arras*, the last two being in the extreme east of that Riding.

One interesting sidelight has been recently thrown upon these Norse invasions by the explanation recently advanced by Zachrisson[1] of the name *Birkby*, found in Allerton wapentake, which goes with *Birkby* near Leeds, a *Birkby* in Lancashire, and another in Cumberland, *Bretby* in Derbyshire and *Brettargh* in Lancashire. All these contain ON *Breta*, gen. plural of *Brøtar*, used by Scandinavian writers of the British Celts. These were doubtless Britons from north-western England who joined the Viking raiders, and gave their names to the settlements in question.

At the beginning of this lecture stress was laid on the need for co-operation in these matters between the historian, the archaeologist, and the place-name student. The results in this matter have been particularly happy in the North Riding. The place-name student has been able to confirm and at the same time very extensively to supplement the inferences which can be drawn from the historical texts. His conclusions in their turn agree closely with the archaeological evidence as interpreted by Collingwood in his study of the ornamentation of Scandinavian crosses in Yorkshire.

On the question of the intensity of the Scandinavian settlement in the various parts of the Riding and the relation of the newcomers to their English predecessors there is no time to say much. The picture of these matters as reflected in place-names is of course entirely different from that found in any other of the four Danelaw counties

[1] *Romans, Kelts and Saxons*, 46–7.

under consideration. The large proportion of names of purely Scandinavian origin, including such types as *Loskay*, formed from the Norse phrase *lopt í skógi*, "farm in the wood"; Scandinavian inflexional forms such as *Melmerby* in Halikeld from ON *málmar-by*, "farm of sandy ground," *Whenby* from ON *kvenna-by*, "women's farm"; still more such hybrid inflexional forms as *Osmotherley* from the genitive singular of ON *Ásmundr* and the English *lēah*; the evidence for the Scandinavianising of earlier English names, such as *Rawcliffe* in Skelton, which in Simeon of Durham shows the English form *rēade clif*, "red cliff," but later always shows forms like *Roudeclife* from ON *rauðr*, or *Stonegrave*, which seems once to have had a name *Stāninga-grāfa* of pure English form and later became *Staingrif*; the fluctuation in such names as *Newsham* in Ryedale between forms showing ON *nýr* and OE *níwe*, all show how widespread was the use of Scandinavian speech in the county and how intimate were the relations in speech of the English and Scandinavian population in the 11th and 12th centuries, leading ultimately to the development of a dialect which can best be described as Anglo-Scandinavian. The full significance of all this material can only be realised when Lindkvist has completed his great study of Middle English place-names of Scandinavian origin or the Place-name Survey has made a study of the other Ridings and the Danelaw generally and can bring this material into fruitful comparison with that for Lancashire, already gathered by Ekwall, and Cumberland and Westmorland, of which as yet we know only the skeleton outlines.

Finally, something may be said of the Norman-French settlement as reflected in our place-names. Here the evidence is different both in quality and quantity from that which may be gathered for the earlier settlements. The Norman-French conquest of the country was a conquest by a military minority. There was no settlement *en masse* and the Normans maintained their authority by their prominence among the great feudal tenants, their hold over the Church and their occupancy of all the offices of importance in the State at large. With the machinery of Church and State in their hands they set a deep impress on our names as a whole, not by planting new ones among us but by extensive, if unconscious, alteration of those already in use. The only language with which they were at all familiar, at least for the first half-century or so, was French, and in committing unfamiliar English names to writing they were again and again guilty of solecisms which affected not only the written form of the name but often in the end its spoken one too. All this has long since been admirably traced out by Zachrisson. It is interesting to see, in the light of the Survey, just how the business works out in various counties.

One finds in every county that it is just those names which stand in closest association with its feudal, ecclesiastical and administrative history which bear most definitely the Norman-French stamp. In Buckinghamshire *Cippenham*, close to the Norman castle of Windsor, shows Anglo-Norman *c*(= *s*) for English *ch*; in the same county the name of the village of *Turweston*, with a wild multiplicity of medieval forms, going back ultimately to late OE *Thurulfestun* probably owes its curious modern form to the nearness of the Norman castle of Brackley. *Odell*, the head of an important barony, goes back to OE *wād-hyll*, "woad-hill," but in medieval documents appears so often in the curious form *Wahull* that this is the name by which the barony is commonly known to historians. *Maulden* in the same county doubtless owes its curious Middle English forms with initial *Meu-*, *Mau-*, *Meau-*, *Meal-* to the fact that it was a possession of the alien priory of St Faith of Longueville. In the North Riding the valley of the Ure or Yore appears in the Anglo-Norman form *Jervaulx* (= Jarvis), with Anglo-French *j* for English *y* and French *vals*, "valley," while the Rye Valley similarly appears as *Rievaulx* (= Rivers), both being important monastic foundations. So also *Helmsley* in Ryedale, the head of a Norman barony, side by side with the more normal forms which have survived in the present name, had a Norman-French form *Ha(u)melak* which survived into the 17th century in the barony of Roos of *Hamlake*. *Malton* contains as its first element OE *mæðel*, "speech," probably because some ancient moot was held there. Its curious medieval forms in *Maal-*, *Meal-*, *Miau-*, *Meau-*, *Meu-* are due either to its being thus a place of some administrative importance or to the presence of an ancient priory there. In Sussex the name of the barony of *Halnaker* from OE *healfan-æcer*, "half-acre," persistently fluctuates in medieval times between the forms *Halnakre* and *Halfnaked*, while close to it are the villages of East and West *Hampnett* with the Norman-French diminutive added to the English *hampton*. Near the Norman castle of Bramber arose the form *Lancing* for *Lanching* or *Lenching*, and under the shadow of the castle of Hastings the name of its ancient port, *burhwara-hȳð*, "citizens' harbour," underwent its strange transformation to *Bulverhythe*. Finally, not far from Arundel is the village of *Madehurst*. Side by side with medieval forms *Madhurst*, *Medhurst*, which account for the modern form (really pronounced with a short vowel), we have even earlier forms *Meslirs*, *Medliers*. These are Norman-French forms for OE *mæþel-hyrst*, "speech-hill." The prevalence of these forms is probably explained by the fact that at the north end of Madehurst parish is "No Man's land," known to have been the early meeting place of the sheriff's tourn for the rapes of Chichester and Arundel. Official usage gave rise to these strangely corrupt forms.

Such names as are definitely of French origin can usually be explained in similar fashion. There is a *Broyle*, an old French term for a park or warren, near Chichester, and another in Ringmer, a peculiar of the Archbishop of Canterbury. In Ringmer, and again near Battle Abbey, there are examples of *Plashet*, an old French term for a woodland enclosure. Near the Norman castle of Arundel we have the *Rewell*, from OFr *ruelle*, "road" or "track," and not far from the old castle of Knepp is *Hatterell*, containing OFr *haterel*, "crown of the head." The only known parallel to this name is *Hatterall* Hill in Monmouthshire, which doubtless owes its French name to its nearness to Llanthony Abbey. *Marlpost* from OFr *malrepas*, "ill-feeding," belonged to the nuns of Rusper. In Worcestershire *Callans* Wood containing NFr *calange*, "challenge," "claim," a hybrid parallel to the common English *Threapwood*, "chiding wood," is close to the abbey of Pershore. *Beamond* in Buckinghamshire is hard by the abbey of Missenden, *Beaupre* in Huntingdonshire is in the *banlieue* of the abbey of Ramsey. There are very few place-names of French origin for the existence of which it is not thus possible to find specific cause.

One further illustration of Norman-French influence is not without significance. Of all the various place-names which give difficulty in interpreting, there is one group which in almost every county gives special difficulty, viz. the hundred and wapentake-names. The difficulty arises in part of course from the fact that the sites of the hundred-meeting places are often unknown, and local topography cannot in such cases help us in interpreting the forms, but the difficulty may arise even where the hundred-name survives as a place-name of some importance. The real difficulty lies as a rule in the widely variant forms in which such names have come down to us. The hundred-name was essentially an official name and the officials have done their best—or worst—with them. A couple of illustrations must suffice. The name of the Sussex hundred of *Dumpford* appears in the following forms in Middle English documents: *Hamesford, Emmedeford, Dyneford, Dymford, Demesford, Demetford, Den(e)ford, Demeford, Dempford.* It is only after some time and thought that one realises that this is probably from an OE *dæmmede-ford,* "dammed ford," *i.e.* one by which there was a dam for a mill-stream. Or why does the Sussex hundred of *Flexborough* persistently fluctuate between the forms *Flexeberge, Fexeberge, Faxeberg, Flaxberghe* when there is no known phonological principle to account for it? It can only be accounted for as another example of the strange vagaries of the Norman or French official mind when it had to deal with English place-names.

LECTURE II

The Vocabulary of our Forefathers

No systematic use has hitherto been made of place-name material as a means of enabling us to recover the vocabulary of our forefathers at any stage in their history. In the later stages of Toller's revision and completion of Bosworth's *Anglo-Saxon Dictionary* the value from this point of view of the material to be found in the boundaries set forth in Anglo-Saxon charters was first fully realised, but little or no use was made even then of Anglo-Saxon place-names not found in such lists of boundaries, and place-names as such play little part in the material used in the *Oxford English Dictionary*. The reasons for this apparent neglect are obvious. When these dictionaries were compiled the material had not been collected, still less analysed and presented in a form which would make it serviceable to the lexicographer. But there is a more fundamental reason. The place-name or the place-name element has as a rule no context which the lexicographer can use for determining its meaning, and even when it is found in a piece of continuous topographical prose, such as a list of boundaries, there is not that necessary logical sequence of thought which enables him to argue from the context what the meaning of the term may be. The significance of any word found in such a list can only appear when the boundaries are studied, either through maps or on the ground itself, in relation to what they actually describe. An interesting illustration of this is to be found in the history of the word *streat* which, on the basis of charter material, sprang into momentary existence a few years ago. The word occurs two or three times and looked as if it might be a word denoting *bushes, thicket* or the like, standing in ablaut relation to the word found in MHG *striuze,* "thicket, copse," but examination of the topography of the charters in which it was found showed at once that *streat* was only a bad form for *stræt,* for just where you get to the *streat,* when you follow the boundaries on the map you find a Roman road[1]. In recent years much has been done by Grundy and others in remedying these and other similar dangers necessarily attending a use of charter material which has not been brought into relation with the ground which it describes.

The task of detecting a new place-name element, still more that of interpreting it when detected, is not an easy one. There are some elements long since recognised and of

[1] For the reference *v.* Bosworth-Toller, *Supplement, s.v.* The corresponding adjective *streatan,* supposed to be found in *on ða streatan hlywan,* disappeared in a different way when *streatan,* as first guessed by Bradley, was ultimately shown to be an error of transcription for *greatan,* "great."

fairly frequent occurrence, of whose exact sense we are still uncertain. The meaning of the element *dræg* in the numerous *Draytons*, *Draycots*, and a few other compounds, is still unknown. There is fair agreement that it is connected with OE *dragan*, "to draw," and denotes a place where the action of dragging took place, but it is difficult to find any common factor in the topography of all the *Dray*-names which will show clearly just what was "drawn" at that place. Similarly with the omnipresent names ending in *-hale*, while it is clear that in many cases the rendering of OE *healh* as "nook, corner" fits the case, it is at times difficult to feel that it necessarily means anything so precise as that, and one is inclined to be as general in one's interpretation of it as Simeon of Durham was when he speaks of *Hearrahalch, quod interpretari potest locus Dominorum*, and makes it mean simply "place."

Great though the difficulties may be in arriving at the precise significance of place-name material, there is no doubt about its extreme importance, not only for the lexicographer but for the historian of our language, race, and civilisation as a whole. For the earlier periods of our history, our knowledge of the secular vocabulary, the language of everyday life, is all too scanty. The literature is largely religious and homiletic in character and has its obvious linguistic limitations. Much can be learned from glosses and vocabularies, but it is clear that there will still be great gaps in our knowledge, and help from place-names, which in their origin and history touch so many different activities and interests in the daily life of the past, ought to be of service in filling some of them. There is no doubt that this is the case, but before proceeding to demonstrate it a word or two should perhaps be said in illustration of some of the various ways in which new place-name elements come to light and old ones are interpreted.

Occasionally a medieval writer himself lightens our task by interpreting some difficult or strange element for us, but he is seldom so illuminating as is Reginald of Durham when, in referring to *Plumbland* in Cumberland, earlier *Plumbelund*, he tells us that *lund* means *nemus paci donatum*, "grove dedicated to peace," and gives us an interpretation which jumps very happily with the known use of *lundr* for "a heathen grove" in Scandinavia itself[1]. More often he is baffling, after the fashion of Bede's interpretation of *Streonæshalch*, the old name of Whitby, as *sinus fari*, or obviously ignorant of the elements of linguistic knowledge, as when the writer of a life of St Oswald tells us that Ramsey in Huntingdonshire is so called from the Latin *ramis*, "branches," and the English *ig*, "island," for "the island is as it were hedged around by great trees[2]."

Equally rare are the occasions on which the topography of a place-name containing such an element is of so striking and definite a character that, when taken in conjunction

[1] *v.* EPNS iii, 220 and note.

[2] *Historians of the Church of York*, i, 432.

with the early forms of the name, its meaning at once becomes obvious. The first element *Totern-* in *Totternhoe* in Bedfordshire remained a mystery until one read the statement in the Victoria History of that county, with reference to the ancient camp there, that "to those moving on the lower plains for miles round, the Totternhoe mound seems to keep watch on its height like some great conning-tower." Then one realised that this must be an unrecorded OE **tot-ærn,* a "toot" or "look-out" house, aptly applied to such an earthwork. Similarly when in Worcestershire one came across three examples of OE *weard-setl* on the tops of prominent hills, with an earthwork still remaining in one case, it was clear that one had to do with another term, actually on record, meaning "watch-seat," which might also be applied to a work of this kind.

At times a suggestion put forward in the first instance on philological grounds has proved on topographical investigation to be specially apt. Ekwall suggested that *Great Whyte,* the name of a street in Ramsey in Huntingdonshire, with ME forms *la Wihte, le Wyghte* and the like, might go back to an OE **wiht,* "turn, bend," from the stem of *wīcan,* "to give way." It was discovered afterwards that the street covered an old waterway which had a bend in its course. Since then one has noted *Wetmoor* in Staffordshire[1], with the same first element, lying in a bend on the Trent; *Whitehall* in Tackley in Oxfordshire

[1] *v.* Duignan, *PN of Staffordshire, s.v.*

(*Wihthull* in KCD 709), lying in a well-curved hollow; and *White* Place by *Whitebrook* in Cookham in Berkshire[1], with ME forms as for the Huntingdonshire name, the brook itself making a semicircular bend at this point.

Sometimes careful comparison of the topography of one or more places containing a certain element readily suggests a word in some other Germanic dialect which would describe them, and at the same time affords an apt philological parallel to the word whose history we are seeking. *Hodshrove* and *Hoadsherf* in Falmer and in Cuckfield in Sussex lie respectively at the foot of a steep slope of the downs and on the side of a hill in a hollow and have ME forms in *schorf*, *schorve*. These at once suggest ModGer *schroff*, "abrupt," and clearly give us a cognate of MHG *schorf*, *schroff*, of which the general sense is "cliff, bluff."

Similarly *Shoreham* in Sussex lies at the foot of the downs, *Shoreham* in Kent (*Scorham* in BCS 822) at the foot of a steep hill, *Shorwell* in the Isle of Wight in a deep valley at the foot of the downs, while *Waldershare* in Kent (*Waldmeresscora* in BSC 381) lies in a hollow of the downs. This clearly suggests association with OHG *scorro*, ModGer *schor*(*r*), *schorre* (dialectal), "steep declivity." The word perhaps still survives in the Scottish *shore*, "steep rock," and is connected with OE *scorian*, "to project."

At times the element or elements composing a place-name are sufficiently obvious, but only patient study of all

[1] 1342 Close Roll.

the examples of its occurrence will reveal its exact significance. There are a number of places in England called *Forty*, the name being specially common now, and still more so in medieval times, in Worcestershire and Gloucestershire. The early forms show that most of them come from OE *forð*, "forth, forward" and *ēg*, "island." The meaning of this somewhat strange compound only became evident after careful topographical examination of all the examples by Mr Houghton. It is a term used of "an island or peninsula of land standing well out from surrounding marshy or low-lying land."

Sometimes a stray hint in the document itself is of service. The term *brook* is used in Sussex of a "water-meadow," "marshy ground." When in a Battle Abbey deed we have a grant of *la Sneppe in Broco* we have clear confirmation that OE *snæp*, ME *snappe, sneppe*, of fairly frequent occurrence in Sussex place-names, is the dialect word *snape*, "boggy ground," hitherto only recorded for Dorset, Somerset and Devon.

Philology and documentary evidence may combine in clearing up the history of some word of uncertain or disputed meaning. *Cockshot* or *Cockshoot* is very common in place-names and is commonly explained as a "glade in a wood through which cocks might dart or 'shoot,' so as to be caught by a net stretched across the opening." Zachrisson showed on the evidence of place-name forms that the word had nothing to do with the verb "to shoot" but contained an OE *sciete*, "corner," so that the name meant simply "corner of land frequented by cocks[1]." Just at the same time Mr St Clair Baddeley showed that the cocks might well be woodcocks, for he found a 15th century deed in which *Cockshoot* Wood in Painswick was held by a rent of "two woodcocks."

Turning now to a more general survey and analysis of the various additions to our knowledge of the vocabulary of the past which may be made in this way, we may begin by noting how many words, often of obscure or difficult origin, can with the aid of place-name material be carried back far earlier than the dictionary material would suggest. The word *ferry* is first recorded as a noun in the *Promptorium Parvulorum* in 1440, but there was a *Blanchefery*, apparently a white ferry-boat, across the Nen already in 1279. The common term *spurt* or *spirt* for a "jet of water" is not recorded before 1716 and even the verb from which it is derived is not on record before 1570, but we have *la Sperte* in 1320 referring to a still existing spring in *Spurt* Street in Cuddington in Buckinghamshire, and almost certain examples of the same word in Worcestershire going back to 1249. The *pewit* that haunted *Pewytelowe* in Cleeve Prior in the 13th century was there three hundred years before Skelton summoned the *pewyt* to sing versicles in *Philip Sparowe*, and you could speak of a *plat* of ground in Lancashire[2] three hundred years before the term was used

[1] ZONF 2, 146. [2] *PN of Lancashire*, 31 and note.

in the 16th century *Domesday of Inclosures*. The word *crab* used of the fruit of the wild apple tree is not on record before 1420, but there is a series of place-names, *Crabbet* in Sussex, earlier *Crabbewyk*, *Crabble* in Kent, earlier *Crabbehole*, *Crabwall* in Cheshire, earlier *Crabbewalle*, going back to the beginning of the 13th century, which must contain this word, and will serve at least to show that one of the etymologies suggested for this difficult word, viz. association with Swedish *skrabbe*, with loss of initial *s*, is unlikely.

Sometimes it is only a case of carrying back a specialised use of a word and not the word itself. *tail*, in the sense "lower end of a pool or stream," is first recorded in 1533, but in connexion with *Tail* in Crewkerne in Somersetshire we have a reference in 1292 to *Taile* water-mill[1], clearly involving this use of the word.

These are common words, still in general use, but the history of many rare or dialect words is often made clearer:—

catsbrain is used in the dialects of Surrey, Sussex, Staffordshire and Shropshire of "a mixture of clay and chalk soil" or of "rough clay mixed with stones." This curious term, embodying some piece of obscure and forgotten animal-lore, has been found in the form *Catesbragen* in the 12th-century cartulary of St Nicholas' Hospital, Salisbury, as *Cattesbreyne* in 1348 in Buckinghamshire, still surviving as *Catsbrain* in Oakley; and there are other early examples in Bedfordshire.

[1] 1292 Ipm.

queach is a dialect term for a "thicket," our earliest reference for which belongs to the 15th century, but we have *la Queche* in Worcestershire in 1307 and it is found in *Quecheworth*, now Cotchford, in Sussex in 1274. Since it is compounded with *worth*, it is probably far older than the bare date would suggest.

quab or *quob* is used in dialect of a "marshy spot, a bog," and is first recorded by Minsheu in 1607, but it is clearly to be found in *Quob* in Hampshire, *la Quabbe* in 1307, *Quobwell* in Wiltshire, *Cuabbavella* in 1292, *Quabbrook* in Sussex, *Quabbalke* in 1285; and even more important, in the phrase *on heahstanes quabben*, *i.e.* "in Heahstan's marshy lands," in a 10th-century Dorsetshire charter, unfortunately only preserved in a Middle English text, but whose genuineness as a record of Old English vocabulary we have no reason to suspect[1].

cleat, a dialect name for "colts-foot," is used by Trevisa in 1398 of "burdock" but is found in the 13th century in *Clethill* in Worcestershire and already in the 12th century in a lost *Clethale* in the same county. It goes back to OE *clǣte* which must have existed side by side with the common *clāte*, "clote, burdock."

brame, recorded from Lincolnshire and Westmorland as a dialect term for "bramble," is found already in the 12th century in *Brampton* in Huntingdonshire[2].

[1] 1292 Ipm, 1307 Ipm, BCS 1218.
[2] ASC *s.a.* 1121.

The rare dialect word *risp*, "bush, branch, twig," only recorded from East Anglia, is probably to be found in *Ripsley* and *Ripshook* in Sussex, for the former of which we have a form *Ripselye* as early as 1265. This must be the same word as ModGer *rispe*, "shrubs, briers," which Falk and Torp[1] believe to show just the same metathesis from earlier *ripse* which we have in the dialect word.

Sometimes place-name material enables us to carry back a dialect word to some earlier and less specialised usage. *crome*, *cromb* is a common East Anglian dialect word, already found in the *Promptorium* and used of a "hook" or "crook." Smith has shown that this word, in the dative plural from *Crambom*, is found already in Domesday in the form in that document for *Crambe* in the North Riding, where it refers to the serpentine bends of the Derwent at this point.

Sometimes place-name evidence compels us to revise our views as to the ultimate history of a word. *polder*, "low-lying land reclaimed from the sea," is first used of English soil by Somner, the Kentish antiquary (*c.* 1609), who says that our ancestors used it of a "marish fen." The OED takes the word to be from the Dutch *polder*, MDu *polre*, but Somner was certainly more correct in speaking of its use by our ancestors, for it is not a loan-word from the Dutch but a native English one, as can been seen from the evidence of place-names. It is the same as *Polders* in

[1] *Etymologisk Ordbog*, *s.v.*

Woodnesborough in Kent, *Poldre* in 1232 and *Polre* in 1246, and is used in *werklond vocata Polre* in Fleet in Lincolnshire in 1316[1], and in *le Newpolder, the polder* in Playden in Sussex in 1404. All of these are in marshland and the term is clearly only the cognate of the Dutch word, not derived from it.

The words with which we have so far dealt do not, so far as we know, except in the case of *quab* and *cromb*, go back to the Old English period. We may next note examples of words on record in Old English, surviving in Modern English dialects, but whose history in the intervening stage has hitherto been largely a blank. Here it will be useful to distinguish words on independent record in Old English and words found only in charter material. Of the former we may note the following:—OE *stearn*, "sea-swallow, tern," gives rise to the Norfolk name *starn* for the common tern, but is also found in *Starnash* in Sussex, earlier *Sternerse*, "tern-haunted stubble-land," not far from the coast, lying low by the Cuckmere river.

OE *heolstor*, "hiding-place, retreat," is the source of the word *holster*, *hulster*, used in Somerset, Devon and Cornwall of a "hiding-place," and has come to light again in *Hylters* in West Dean in Sussex in the form *la Hulstre* found in 1310. *Hylters* lies in a well-marked valley.

Old English *telga*, "branch, twig," has no history in the dictionaries between the Old English period and the

[1] 1232 Cl, 1246 Ch, 1316 *Terrier of Fleet*, ed. Neilson.

modern dialectal *tellow* (Sussex), *telly* (Yorkshire), used respectively of "a young sapling" and "a lateral shoot from a stalk of corn." The history of *Sweetwillow* and *Tilkhurst* in Sussex, from OE *swēte telga* and *telga-hyrst*, illustrates the use of this word in that county, and the ME forms of the names show clearly that the OED is right in surmising that the forms *tellow* and *telly* may go back to OE *telga*, whereas the more common variant *tiller* clearly goes back to OE *telgor*.

The rare Old English word *blēat*, "bare, miserable," once recorded in Middle English in *The Owl and the Nightingale*, is found in the dialect word *bleat* recorded from Gloucestershire, Wiltshire, Surrey, Sussex, and Kent, but only known elsewhere in the Sussex *Bleatham*, an old name for the parish of Egdean, which goes back to the 12th century.

We now turn to those cases in which our only early authority for a dialect word is some word preserved in charter material. Two of these, *shore* and *snape*, have already been dealt with in a different connexion[1]. We may note a few more:—

ripple, "coppice, thicket," is recorded from Herefordshire, and is found frequently in the form *rippel* in Old English charters[2]. It survives in *Ripple* in Worcestershire and *Ripple* in Kent. The root-idea is however "strip," as in

[1] 46–7 *supra*.

[2] Bosworth-Toller, *Supplement*, *s.v.*

Norwegian *ripel*, "strip," the sense "strip of wood, coppice," being only a secondary development. The two place-names perhaps bear witness to this, for both the Ripples are on projecting spits of land, and their topography suggests the earlier usage.

There is a dialect word *snook*, used in Northumberland and Somerset of a sharp-pointed projection. Our sole earlier authority for it, collected in this case in the OED, is place-name material, but even there it is only carried back to the 13th century. It is actually found in the 10th century in a Somersetshire charter (BCS 959) in the form *snocan*, which may be either the dative singular of a weak form *snoca*, or the dative plural (late) of a strong or weak form *snoc*(*a*). We also have the form *westsnok* in another Anglo-Saxon charter (BCS 1313), unfortunately only preserved in a ME text.

The etymology of the word *thurrock*, used in Kent and elsewhere of "a drain," is obscure (*v.* OED *s.v.*). The earlier history of the word, first guessed by Bradley, is in part at least cleared up by the place-name forms for *Rockmoor* in Hampshire, which are *þrocmore, þorcmore, þorocmere* (BCS 508, 1080). These suggest that *thurrock* may be for earlier *þor*(*o*)*c*, a metathesised form of *þroc*. The fact that the element *þrocc* is associated with *mere*, "mere," in *Rockmoor*, with *mor*, "marshland," in *Throckmorton* in Worcestershire, with *mill* in *Drockmill* in Sussex, and with OE *bæc*, "stream," in a lost *Throkbach* in Worcestershire, tends to confirm the association of OE *þroc* and the dialectal *thurrock*[1].

In an original 8th-century charter (BCS 160) we have mention of *silbam qui appellatur Ripp*, somewhere near the borders of Sussex and Kent. This may actually refer to *Ripe* in Sussex, for which we have early forms *Rip, Ripp, Ryppe.* Even if it does not refer to that place itself it is clear that the names are identical and that we have only a second example of the use of the term *ripp.* The topography of *Ripe* is that it lies on a small but well-marked ridge rising above marshy ground. The same element is found in *Ripton* in Huntingdonshire, with a lost *Ripthornes* close by. These are in an old woodland area on relatively high ground which, two miles to the north, sinks down to the fen-level. The element *ripp* or *ripe* in these names is clearly the Kent and Sussex dialect word *ripe* meaning "shore, bank," which must be the cognate of LGer *riep,* "shore, slope," East Frisian *ripe,* "edge."

Finally under this head we may note the new light thrown on the history of the word *sangle* used in Devon and Cornwall and *songle,* used in Cheshire, Shropshire and Herefordshire, of a "handful of gleaned corn," which apparently takes the form *single* in Scotland. This must be the word *sængel* found in the Sussex place-name *Sængel-*

[1] There is also an OE word *þrocc,* "timber to which the ploughshare is attached" (dial. *throck*), but it is difficult to see how this could be found in place-names.

wicos (sic) in BCS 144, perhaps in the neighbourhood of *Singleton*, which in its ME forms varies between *Sangelton* and *Sengelton*, and in *la Sengle*, the medieval name for part of the present Cowdray Park, possibly also in the Sussex *Chancton*. This would seem to be the same word as LGer *sangele*, used in Westphalia of "a small bundle," "tuft," and found by Jellinghaus (*Die Westfälischen Ortsnamen, s.v.*) in various place-names, an expansion of the simple *sange* used in the same sense. It looks as if the word, in addition to the sense recorded in the West Country, could also be used to denote a "bundle" or possibly a "thicket of brushwood," for both the Sussex places are in old woodland areas.

All these OE words have left traces of themselves in modern speech, whether standard or dialectal, quite apart from the evidence of place-names, but there are a good many OE words which have left no trace at all in the later language but can be found still surviving in fossil-fashion in our place-names.

In the fen district in Huntingdonshire, near Whittlesey Mere, we have *Chalderbeach* and *Rawerholt*, the latter no longer on the map, which preserve the OE names *scealfor* and *hragra* used respectively of a species of diver-bird and of the heron. *Rat* Farm in Battle in Sussex and *Rutt* in Ugborough in Devon, with ME forms *Rette, Rutte, Rytte,* preserve the otherwise unknown OE *ryt*, "rubbish for burning." *Lent* in Taplow in Buckinghamshire, with ME forms *atte Lenthe* or *Liente*, is a survival of OE *hlēonaþ*, "shelter," a derivative of OE *hlēo*, "lee, shelter, protection." OE *lǣl(a)*, "twig, branch," does not seem to be known in post-conquest times apart from the place-name *Lealholme* in the North Riding, which comes from its dative plural form, and the same is true of OE *clūse*, "enclosure, narrow passage," and the like, which survives in *Clows* Top near Mamble in Worcestershire and in *Clowes* in Blean in Kent. The very rare OE *hlēde, hlȳde*, "seat, bench," used perhaps as a farm-name after the same fashion as OE *setl*, survives apparently in *Lude* in Wooburn in Buckinghamshire. OE *þille*, "flooring, planking," is only known in *Dill* hundred in Sussex and *Dylle*, the name of a tithing in Foxearle hundred in the same county. The early forms clearly go back to OE *þille*, the reference being probably to some temporary wooden structure used at the hundred-moot.

An interesting word of this type is the OE *medeme, medume*, "small, moderate, middling." This survives in *Medmenham* in Buckinghamshire and in *Medmerry* in the Selsey peninsula in Sussex. In these names, as Ekwall suggests, we may have the earlier and more primitive sense of "middle," *Medmenham* being the middle one of three *hamms* on the river or the middle *ham* between *Bisham* and *Remenham*, while *Medmerry* lies between *Thorney* and *Selsey*. We have already noticed the curious survival of OE *būrscipe*, "community of peasants" in the Sussex *Boship*. Of the same type is the survival of OE *gemǣnnes* in numerous ME examples in Sussex, in *Minnis* in Hastings at the present day and in the word *mennys* used in Kent of a large common. The old English word denoted in the first instance "joint or common tenancy of property," but was later transferred to the property itself, cf. BCS 426, "in commune silfa quod nos saxonice *ingemennisse* dicimus."

Sometimes the fossils thus preserved are of extreme age, clearly going back to the first days of the settlement. The word *dryhten*, "lord," is used in the heroic poetry of a king or prince and became part of the stock poetic vocabulary, as when Aethelstan is called *eorla dryhten* in the *Battle of Brunanburh.* It is also freely used in OE poetry of Christ and God, this tradition continuing into Middle English. It is never found however in OE prose, and when the written literature begins was clearly an archaic term such as could only be employed in poetry, yet we find it in *Drigsell* in Salehurst in Sussex, which is clearly from OE *dryhtenes-(ge)selle*, "lord's hall," and it is found also in *drihtnes dene* in an 8th-century Gloucestershire charter (BCS 166).

Of a similar archaic type is *Earith* in Huntingdonshire, the same name as *Erith* in Kent, for which we have a form *Earhyð* in a copy of a 7th-century charter (BCS 87). The first element in this name can only be the word *ēar*, for which our sole record is the *Runic Poem*, where it is the name of one of the runic letters. In the place-names in question it probably describes a muddy *hythe* or landing-place, and has the sense of its ON cognate *aurr*, "wet clay" or "loam."

Finally under this head we may note an interesting relic of Old English heathendom. There is an OE word *wīg*, *wīh*, *wēoh* used in the Christian poems of the idols of the heathen. In the Gnomic Verses of the Exeter Book we are further told that "books are for the learner, housel for the holy man, sins for the heathen, *Woden worhte weos*." *weos* is ambiguous in form; it may be either the genitive singular of *weoh*, involving a somewhat rare construction of *wyrcean* with a genitive object, or the accusative plural of the same word, so that Woden may be credited with having made one or possibly more *weohs*. All these are examples of Christian writers reproaching the heathen for their use of *weohs*. When we turn to place-names we can find several cases of heathen settlers using the term in connexion with their own heathen worship. *Weedon* in Buckinghamshire and *Weedon* in Northamptonshire go back to OE *Weodun* (KCD 824, BCS 792), *Willey* in Surrey to *Weoleage* (BCS 627), *Whiligh* in Sussex to OE *Wiglege* (Stowe Charters 38). *Whyly* in Sussex, *Weoley* in Worcestershire and *Weeley* in Essex, though not found in OE documents, clearly go back to the same combination of elements as these last. Still more interesting is *Patchway* in Falmer (Sussex). The name only survives locally, but a chain of forms with persistent final *-wy*(*e*) carry it back to the corrupt *Petteleswig* found in the very bad text of an 8th-century charter (BCS 197). Here we have the *wig* or *weoh* owned by one *Pæccel*. Similarly we have reference in a 7th-century Surrey charter (BCS 72) to a *Cusanweoh*, a *weoh* owned by one *Cusa*. The question arises, "What exactly is intended by a *weoh* in these various names?" It is not at all likely that they were idols. Such played little part in Teutonic heathendom and it is more probable that the word *weoh*, the root-idea of which is simply "sacred, holy," is here used of some sacred place, possibly even of a rudimentary fane or temple, like its Norse cognate *vé*. The favourite site for such heathen fanes would seem to have been a hill-top or a forest grove or clearing, to judge by the elements *dun* and *leah* with which *weoh* is compounded. This agrees well with the use of *hearg*, the only other OE heathen term for a sacred place with which we are familiar. In *Harrowden* (Bedfordshire and Northamptonshire) it is compounded with *dun*, "hill." *Harrowbank* (co. Durham) is, as the name implies, a hill. We all know that *Harrow* in Middlesex is "on the Hill," and *Peperharrow* in Surrey is similarly on a hill above the Wey. *Arrowfield Top* in Alvechurch (Worcestershire) also contains this element. We may note also that of the four place-names known to contain the name of the god *Woden*, three, viz. *Woodnesborough* in Kent, *Wednesbury* in Staffordshire, and *Wenslow* in Bedfordshire contain the OE words *beorg* and *hlaw* denoting a hill, and the only place that can be definitely associated with the worship of his Scandinavian counterpart is the remarkable rounded hill in Cleveland, now known as Roseberry Topping but formerly called *Oþenesberg*. The only other heathen Teutonic gods as yet noted in place-names are *Tiw*, the god of war, and *Thunor*, the thunder-god. The former is found in a compound of *leah* in *Tuesley* (Surrey), the latter in *Thundridge* (Hertfordshire) compounded with *hrycg*, "hill," and compounded with *leah* in *Thursley* (Surrey) and in *Thunderley* and *Thundersley* in Essex. The only other element that has been found compounded with these heathen god names is *feld*, as in *Wednesfield* (Staffordshire) and two examples of *Thundersfield* in Surrey, where *feld* denotes "open country," possibly a larger opening in woodland than a *leah*. In addition to these we have unidentified examples of *þunoresleah* in Hampshire and Sussex (BCS 393, 208), *þunoresfeld* in Wiltshire (BCS 469), a *Wodnesbeorg* in Wiltshire (BCS 390) and a *Wodnesdene* in the same county (BCS 734).

Finally on this point it may be worth noticing how intensely heathen south-western Surrey must have been. In the area between Farnham and Guildford we have Willey, Thursley, Tuesley, Peper Harrow, one of the Thundersfields, and the unidentified *Cusanweoh*, which was in the neighbourhood of Farnham. There is no other area in England so thickly covered with heathen names.

After this somewhat lengthy but perhaps not unjustifiable digression, we may pass to the consideration of the somewhat numerous examples of place-name elements not found, or but rarely found, in Old English documents, which are cognates or derivatives of well-established words, but have no history apart from their occurrence in place-name forms. Examples of these are as follows:—

OE *cnoll*, "knoll," is well established, but side by side with it there was an OE **cnyll*, a *jo*-stem, which has given rise to *Knell* in Goring and *Knelle* in Beckley in Sussex, *Nill* in Cambridgeshire, *Woodknowle* in Devon, names which show ME forms in *Cnelle-*, *Cnolle-*, *Knylle-*, *Knille-*.

The phonology of *pinfold* gave trouble to the editors of the OED. Side by side with the correct *punfold* and the like, from OE *pundfald*, one gets from about 1400 onwards *pyndefold*, *pynfold*. It is suggested that this may be due to association with the verb *pin* or *pynd* from OE *pyndan*, but it may well be that there was an alternative form of the compound formed at an early date from a noun **pynd*, for that word can be amply established on the basis of place-name evidence. We have it already in *frodeshamespend*, *flothamespynd* in Kentish charters (BCS 335, 336) and the ME forms show that we also have it in *Pen Hill* in Lancing and *Pinland* in West Grinstead in Sussex, in *Piend* in Jacobstow and Stockleigh English in Devon, in *Pendell* in Surrey, to mention only a few examples.

OE *snǣd*, "piece, morsel, bit," is a fairly common word. It is only in OE charters and in place-name material generally that we get evidence, and abundance of it, for the unmutated form *snād*. Middendorff has numerous examples from the charters; we may note its survival in *Snoad* and *Snodhurst* in Kent and in *Snathurst* in Brede in Sussex. It was used of a cutting made into a wood or of an isolated wood.

A similar pair of forms is exemplified in the relation of OE *hǣþ*, "heath," to the element **hāþ*, found in numerous place-names in Sussex and Kent and once in a Kentish charter (BCS 459) in the name *haþdun*. Examples of its survival are the Kentish *Hothfield* and *Hoath* in Patrixbourne and the Sussex *Hoathlys*. In the latter county some eighteen examples of it have been noted; *hawth*, commonly said to be the Sussex dialectal pronunciation of *heath*, goes back not to OE *hǣþ*, but, in strict phonological fashion, to OE *hāþ*. *hǣð*, like Gothic *háiþi* and ON *heiðr*, is a feminine *iō*-stem, while *hāþ* is probably from **haiþa*, a neuter *a*-stem. Dr Mansion has drawn my attention to the interesting Flemish parallel where we have neuter *heed*, side by side with feminine *heede*.

OE *hlinc*, "hill, slope," is familiar to us in *linch* and its derivative *linchet*, still more in the form *link* or *links*. Place-names give us an otherwise unknown cognate. In south-east Worcestershire we have a series of places, Abbot's *Lench*, Rous *Lench*, Church *Lench*, Atch *Lench*, Sheriff's *Lench* and *Lench*-wick, covering an area of some five miles by two, the land being notably broken and hilly. We have late OE forms in *Lenc* for one of them and it is clear that this is for earlier **hlenc*, a derivative not of *hlinc* itself but of **hlank*, another grade of the same stem, found in certain continental Germanic names in the form *lanke*. Ekwall has noted it also in *Lench* in Bury in Lancashire and it survives in the dialectal *lench*, "shelf of rock," etc.

Words denoting a clearing are very common in place-names. On the basis of charter material and place-name forms generally the existence of an OE word *rod*, denoting "a clearing" and surviving in the Yorkshire *royd*, was demonstrated a few years ago. Now we find in Sussex, probably also in Hampshire and the Isle of Wight, certainly also in Kent, a series of place-names containing as their second element a word which appears in the ME forms as *rude, rede*. Examples are *Coldred* in Kent, *Languard* in the Isle of Wight, earlier *Langerude, Langerede*. All these point to an OE **rīed*, **rȳd*, cognate with OHG *riuti*, South German *ried*, "cleared land," standing in the same ablaut-relation to *rod* as ON *rjóðr*, which survives in *Reeth* in Yorkshire, to ON *ruð*.

OE *smygel* is used of a rabbit-burrow. In Sussex we have in the Ambersham charter (BCS 1114) a *smeagel-hyrst* which clearly contains a word related to it. There are also places, now called *Broxmead* and *Brooksmarle*, of which the early forms were *Broksmegl* and the like, and a lost villata of *Smeghell* in the south-east of the county, near Rye. Ekwall very aptly suggested that *Broksmegl* must mean "badger hole" and that in these names as in *smeagelhyrst* we must have an unrecorded **smēagel* standing in ablaut-relation to *smygel*. It is a pity that we do not know the site of *Smeghell* and so cannot find what sort of a "hole" it was.

A slightly different type is the case, again noted by Ekwall, where we can show from place-names an example of a Latin loan-word borrowed in a hitherto unknown form. Late Latin *pruna*, "plum," was brought to England by most of the settlers in the form *plūme*, whence our common word *plum*. The men who settled at *Broomhill* in the south-east of Sussex, which shows persistent *Prum-* in the early forms, must have used a form *prūme*, allied to the LGer *prume*, Du *pruim*.

We may now note a few examples of new derivatives. OE *hylde* is well established. A new compound of it has come to light in *Navant* Hill in Lurgashall (Sussex). This is *Ouelte* in 1327, *Novelt* in 1540. This must be from OE *of-hylde*, "down" or "steep slope," with the common affixing of initial *n* from such a phrase as ME *at then ouelte* and infixing of *n* in the unstressed syllable.

From OE *bærnan*, "to burn," was formed a noun *bærnet*, denoting first the action of burning and then a "place cleared by burning," whence such place-names as *Barnet* in Middlesex. From the verb *sengan*, "to burn, singe," there must, in similar fashion, have been formed a noun **sænget*. This word is frequently found in OE charters, four of them being original charters of the 9th and 10th centuries. It was frequently compounded with *leah* and in the later history of these names was prolific in producing imaginary saints, *St Chloe* in Gloucestershire, *St Clair* in Worcestershire, and, when compounded with *tēag*, "tye, enclosure," *St Ives* in Sussex, the initial ME *Sent-* having been interpreted by popular etymology as a weakened form of *Saint*.

The use of the suffix *-ing*, apparently to denote a collection of things, is illustrated in such names as *Thurning* in Norfolk and Northamptonshire, from OE *þyrne*, "thorn-bush," noted by Ekwall[1]. Other examples of this type of formation have since come to light. Karlström has noted *clætinc* (BCS 1071) from *clāte*, "burdock," or perhaps from *clǣte* (*v. supra* 50). We have a *hæseling* in Sussex, from *hæsel*, "hazel," and *brēmeling* from *brēmel*, "bramble." Another *-ing* derivative is found in *sylling*, from *syllan*, "to give," surviving *atte Sullinge*, a 13th-century personal name in the same county, and forming the first part of *Sullington*, "farm which has been a subject of gift."

Another example of a collective derivative is to be found in *Frant* (Sussex). This goes back to OE *fyrnþe* (BCS 961) and another example of it is found in an unidentified *fernethe* in the same county. These show a derivative of the common *fearn*, "bracken," formed with the suffix *-iþja*.

Of similar interest is the suffix *et*, which is curiously prolific in Sussex. This has more than one source. In the

[1] *PN in -ing*, 15.

Sussex *Grevatts* (twice) and *Gravatts*, both with final *s* of manorial origin, and in *Greatwick*, we have OE *græfet*, *grafet*, "trench, ditch," etc., a derivative of *grafan*, "to dig," of the same kind as that noted under Barnet *supra* 65. There are however a considerable number of other names for which this formation will not account. We have four examples of *Naldrett* and one *Naldred*, three of *Burchetts*, two of *Birchetts*, a *Betchetts*, a *Haslett* and a *Reditts*, all showing ME forms in *-ette* and all clearly derived from tree and plant-names—*alor*, "alder," *bierce*, "birch," *bēce*, "beech," *hæsel*, "hazel," *hrēod*, "reed"—and denoting collections of these. A few examples have been noted in other counties also. This suffix is clearly the Germanic suffix *-itja*, found in collective names of similar formation in Old Dutch. Mansion (*Oud-Gentsche Naamkunde*, 77–8) gives us *Feret* from *ferh*, "fir," "oak," *Fursitium* from *fursi*, "furze," and *Fliterit* from *flither*, "elder." It is difficult to say just how old these names are for they are all attached to quite small places. No single one chances to be mentioned in an Anglo-Saxon charter and none of them occurs in Domesday. We have no means therefore of judging whether formations of this kind were specially common among the South Saxon settlers at the time of their first coming or are a later development.

Another curious but unrecognised use of a common suffix may be noted, even though it is concerned with personal names rather than place-names. In working through the

Subsidy Rolls for Sussex it was noted how very common indeed were second names or surnames ending in *-er*, not only of the agent-type like *Baker*, but such names as *Breggere* or *Bridger* and *Laker*. The former might possibly be interpreted as "builder of bridges" but the latter could hardly be "maker of lakes." The significance of such names first became apparent when side by side with a lost place-name *Rumbrugge* one noted the surname *Rumbridger*, which could hardly mean anything but "person living near *Rumbridge*" and was finally settled when it was noted that a man called Robert *atte Linch* in the roll of 1327 appeared as Robert *Lincher* in that of 1332. Similarly, the family who lived *atte Coumbe* became *Coumber*, one living *atte Hylde* (OE *hylde*, "slope") became *Hilder*, the family of *atte Compe* (OE *comp*, "field") became *Comper*, and a family living *atte Crouch*, *i.e.* at the village cross, was later known as *Croucher*. In Sussex, and so far this type of personal name has not been noted elsewhere, the suffix *-er* could be used then with any topographical element to denote a person who lived by it or had to do with it.

We now turn to a more difficult class of words, viz. those that are found in place-name material alone, not being found in Old or Middle English texts and not having any representative in Modern English whether dialectal or standard. Here, in increasing order of difficulty, we may deal with those words which, though not known in the ordinary texts, are found in Old English charters and those which are found only in post-Conquest place-name forms. The chief difficulty in many of these words is not to establish their philological affinities but to determine with any measure of precision their exact significance.

Examples of the first class are as follows:—

In Sussex we have a place called *Stumbleholm* in Ifield, going back to earlier *Stombelhole* (1327) and there is a *Stumblehole* in Leigh in Surrey, going back to *Stombelhale* in 1332. We have an *Ellesstumble* and a *Stumbelforlonge* in Durrington (13th century) and one Gervas lived *atte Stumble* in the neighbourhood of Hellingly in 1296. These names clearly derive from a lost OE cognate of Ger *stummel*, "stump," which is the cognate of the OHG adjective *stumbal*, "blunted." It is found also in *Stumpshill* in Blackheath (Kent), *Stumbelhylde* in 1332.

Another interesting word of the easier type is the element *slohtre* or *slohtra* found in *Slohtranford* in the boundaries of Salmonsbury in an original 8th-century Gloucestershire charter (BCS 230). Examination of the charter shows that this ford must be by the *Slaughters*, Upper and Lower. These lie in a well-marked river-valley and it is clear that their name is to be associated with LGer *slochter*, denoting "a ditch." This has given rise to more than one example of *Schlochter(n)* as a place-name in Westphalia, and Förstemann gives a few further examples of OGer *sluhter* in place-names. Jellinghaus states that in Westphalia it is used of broken land full of hollows. This element is very difficult to distinguish from *slāh-trēo*, "sloe-tree," and *slāh-þorn* in English place-names, unless we have pre-Conquest forms for the name. *Slaughterford* in Wiltshire[1], though it lies in the deep-cut valley of the Avon, is an example of *slah-þorn*, and it is difficult to be sure about some of the examples of the element *slaughter* in Sussex.

The element *gealt* is found once in a place-name in an OE charter (BCS 261), viz. in *gealtborgsteal*, the name of a piece of woodland in Sussex in an 8th-century charter, only preserved in a late copy. This is not found anywhere else but is of considerable interest as it is the only trace of the existence of the OE cognate **gealt* of ON *gǫltr*, OHG *galza*, "male boar."

In similar fashion we may have a trace of an OE cognate of Langobardic *fereha*, OHG *fereh-eih*, "oak," in the Sussex *Firles*, of which the oldest form is *Firolaland* in an 8th-century charter (BCS 262). The first element here may be the genitive plural of an OE **fierol*, an adjective descriptive of a place overgrown with oaks, later used with substantival force. If Hirt's etymology of Gothic *faírguni*, OE *fyrgen* is correct, we have another derivative of this word for oak in Ferryhill (co. Durham), (*æt*) *Feregenne* in BCS 1256. These names are examples of the extreme archaism of many place-name elements. The old word *ferh* for the oak was early lost from the Germanic languages,

[1] Ekblom, *PN of Wiltshire*, 149.

while the derivative *fyrgen* is only found in OE poetry, never in prose, except for the statement in the Leechdoms *flet Thor on fyrgen hæfde*, "Thor had a dwelling on the wooded hill," which is evidently a piece of proverbial ancient wisdom.

The history of the word which lies behind the familiar *Peak* of Derbyshire is also worthy of note. For this we have forms *Peaclond* in the Saxon Chronicle and *Peacesdæl*, "valley of the Peak," in an 11th-century charter, as noted by the editors of the OED, and the name *Pecsæte* given in the Tribal Hidage (BCS 297) to the people who lived in this district. In the OED the word is said to be "of unknown origin: perhaps British." There is no known British word which will account for it and if we examine the name further, especially in the light of additional place-name material now at our disposal, the history of the name would seem to be different from that suggested there. A further example of *peac* has come to light in a Hampshire charter. In a charter of the Meon district (BCS 758) we have reference to *lytlan weac*. Mr C. A. Seyler has shown good reason to identify this with *Peake* Farms, north-west of Old Winchester Hill, on a prominent spur overlooking the Meon valley. (The form *weac* is due to the common error of transcription of *w* for *p*.) We have also in Devonshire East and West *Peek*, Domesday *Pech*, on a hill rising to 569 feet. In Bedfordshire there is a place called *Pegsdon*. This is *Pechesdone* in Domesday. The later forms make it clear that the second element in this name is *denu*, "valley," rather than *dun*, "hill." The natural thing would be to take the first element in this genitival compound as a personal name, but we have an early reference to a Miles de *Pek*, a tenant in Pegsdon, and as there is a prominent and steep hill just to the south of Pegsdon, it seems almost certain that he took his name from this hill, which may well have been called *peac* in OE days. *Pegsdon* then is one of the rare, but by no means unknown, genitival compounds to be found among our place-names (*v. infra* 97 ff.) and means "the valley of (*i.e.* associated with) Peak hill."

So also in the Bridlington Cartulary in connexion with Ganton (East Riding of Yorkshire), we have mention of *Pec* and *Pekesbru*, *i.e.* "*Pec's* brow," the name still surviving in *Peak* Clumps on a steep hill above Ganton. These names leave no doubt that there was a common word *pēac* in Old English meaning "prominent hill" or the like. That its affinities are Germanic and not Celtic was first suggested by Falk and Torp (*Etymologisk Ordbog, s.v. paak*). There they point out that behind Norwegian dial. *pauk*, "stick, little boy," and MDu *pôk*, "dagger," we have a Germanic stem *pauk-*, cognate with the *pēac* of OE *peaclond*. Ekwall notes that words belonging to this and allied stems often denote something rounded or thick-set and the meaning "knoll, hill," could readily develop from that. In considering the meaning of the term we must beware of being misled by the common English *peak* denoting "something pointed."

The ultimate history of that word itself is very difficult but it is clear that, so far as it is used of a pointed hill, it is so largely because of the influence of the French *pic*. When we speak of the "Peak" of Teneriffe, we are only rendering the French *pic* by a similar sounding word.

Another word found only in charter material and in place-names, but of curiously local distribution, is *glind*, found twice in Sussex charters (BCS 208, 821) in *wican glinde* (properly *pican glinde*) and *glindlea*. It is also found in the place called *Glynde* and in four other place-names, all in South-east Sussex. Middendorff suggested association with MLGer *glind*, "fence, enclosure." Förstemann has examples of *glind* in place-names, going back to the 10th century, so that its early form agrees with that of the Sussex name. We cannot be sure of its meaning in Sussex but it may well have had the same sense as its Low German cognate.

More difficult as to their probable meaning are the three words *clopp*, *spic* and *stiorf*. The first of these is found in *Clapham* (Bedfordshire) in the form *Cloppham* (KCD 809), in *Cloppaham* for *Clapham* (Surrey) in Alfred's will (BCS 553), in *Cloptun* (BCS 1061) for *Clapton* (Nth), and the same form in KCD 666 for *Clopton* (Warwickshire). We also have unidentified *clopæcer* and *clophyrst* in BCS 1282 and *cloptun* in KCD 649, as well as *Clapton* in Middlesex, Berkshire, Cambridgeshire and Suffolk, *Clapcote* in Berkshire and Wiltshire, *Clophill* in Bedfordshire, and two examples of *Clapham* in Sussex. All these contain an unrecorded OE **clopp*. The only known Germanic cognate of this word is Middle Danish *klop*, "block, lump," as noted by Skeat. The probability is that the word *clopp* in OE denoted "tree-stump" or the like, the *ham*, *tun*, or *cot* being in the neighbourhood of such, the *æcer*, *hyrst* and *hyll* having one or more such upon them. *Cloppaham* is curious and difficult; it seems to mean "homestead of the tree-stumps," *i.e.* perhaps "marked by such."

spic is even more difficult. This term is again of somewhat narrow distribution, as so far observed. It is found once in a Kentish charter (BCS 175) in *holan spic*, the name of a swine-pasture, where it is apparently compounded with the adj. *holh*, "hollow." It is found occasionally in ME place-names in Kent and Sussex, but the only ones that have so far been identified are *Poles Pitch* in Shipley (Sussex), which lies in a gently sloping valley, and *Mispies*, down on the Pevensey Levels, where it is compounded with OE *mōr*, "marsh-land." This must be cognate with *spik* used in Westphalia (*v.* Jellinghaus, *Die Westfälischen Ortsnamen*, *s.v.*) of "a pool" and "a fish-weir." The exact sense in Old English must remain obscure.

The same is the case with the element *stiorf*. This is found in a Kentish charter in the name *biddanstiorf* (BCS 502), in a Sussex charter (BCS 208) belonging to Bexhill in the form *steorfan* (probably a dative plural form though possibly a dat. sing. of a weak form *steorfa*) and in three other places, all in South-east Sussex. It is clear that in these names we have an element containing the same root as the common word *steorfan*, "to die," but the exact sense of the term it is impossible to discover. It has been suggested that this is an adaptation of OE *steorfa*, "pestilence," to denote a place where there had been a cattle plague, but it does not seem very likely that such a term could come to be used as a regular place-name, though it might be so used in an odd instance or so.

Last of these rare words found in OE charters we may note the element *bing* found in *binguuellan* (BCS 208) in the Bexhill charter, the site being unknown, and probably also in a *Byng(e)ley* in Heathfield, which may be identical with the site of the present *Bingletts* Wood, but this is by no means certain. There is an English dialectal *bing* used of (*a*) a heap, (*b*) a receptacle of various kinds, but these are North Country words and loans from the Scandinavian dialects. Falk and Torp however (*s.v. bing*) relate the second of these two words to MHG *binge*, "hollow in the hills," and Middendorff (*s.v.*) relates the charter-term to High German *binge*, "forest-ditch," used in place-names. There may have been in Old South Saxon a word *bing* denoting a hollow of some kind. Such an element would readily compound with *well*, and if *Bingletts* is really identical with the old site of *Byng(e)ley*, then it may be noted that it lies in a definite hollow.

This brings us to the end of the new words for which we

have OE forms in the charters. We must now consider a few found only in ME place-name material.

We may first note a few terms which have to do with hills. On the Worcestershire-Staffordshire border there is a group of hills known as the *Clent* Hills. Though this reminds us at once of the well-known Danish *klint*, "rock," as in *Moens Klint*, the form of the name, already found as *Clent* in Domesday, and its use in the West Midlands alone, prevent our deriving it from that word and taking it to be a Scandinavian loan-word. Rather we have a derivative of a Germanic **klant*, standing in ablaut relation to the **klent* from which the Danish term is derived. This form *klant* lies behind place-names in *klunt* in the North Frisian dialects and the modern Norwegian *klant, klatt,* all used with the same sense as *klint.*

So also *Bouts* in Inkberrow, near which we have two or three small isolated hills, contains another lost word for a hill. Jellinghaus notes in Westphalia an element *bolte* or *bult* denoting a small rounded hill. The same word is found in MLGer *bulte,* ModLG *bulte,* Du *bult,* Swiss *bulzi,* used of "something rounded, a heap, a small hill." The ME forms of *Bouts* are *Boltes, Bultes,* and these taken with the topography prove beyond doubt that we have to do with a lost OE **bult,* the cognate of these words, showing preservation of the original *u* owing to the influence of initial *b* and possibly also because of the following *l,* as in some other OE words.

Not quite so clear in its exact sense is the word which lies behind *Harome* (North Riding of Yorkshire) and *Herne* (Bedfordshire). The Domesday form for the former is *Harum*, for the latter the ME forms, not of earlier date than the 12th century, vary between *Hare* and *Haren*. Both alike seem to be dative plural forms from some word *hār* or the like. Ekwall called attention to the cognate Swedish *har*, "stony place," *stenhar*, "heap of stones." Jellinghaus notes several places called *Haar* and *Haaren* in Westphalia, also the use in MLGer of *hare* to denote "hill" and the use in the Netherlands, inferred from place-names, of *haar* to denote a wooded height. Behind *Harome* there rise two or three rounded hills above the valley of the Rye, while *Herne*, which is the name for a considerable area, is broken hilly country. It looks very much as if the word denoted "hill" in this country, too. It is unfortunate that, except when used by itself, it is practically impossible to distinguish it from OE *hara*, "hare," and *hār*, "grey," so that opportunities for detecting its presence and studying its topographical significance are peculiarly limited.

There must have been in OE a lost **slinu*, "slope," related to Norwegian *slein*, "gentle slope," as noted by Ekwall in dealing with *Slyne* in Lancashire. This is found also in *Slindon* in Staffordshire and Sussex.

Another word for "slope" is found in the *slind(e)* of *Slinfold* (Sussex), earlier *Slindefald*. Here we have a word cognate with OSw *slind*, "side," which, to judge by its cognates, probably also originally meant "slope."

One or other of these words lies behind *Slimbridge* (Gloucestershire), *Heslinbruge* in Domesday. Slimbridge lies at the end of a slope going down to low-lying land by the Severn.

From the hill we pass to the valley and we may note the strange word *gill*, so frequent in the Wealden area that it is recognised in the dialects both of Surrey and Kent in the sense "narrow wooded valley." There is a Scandinavian word *gil* with this meaning, familiar to us in the *gill* or *ghyll* of the Lake district and North-western England generally. This word has often been used by lay enthusiasts as an example or as a proof of the influence of the Jutish settlers in this country, forgetful of the fact that this word is not in use in Denmark, but still more of the fact that the *Jutes*, whoever they may have been, did not speak the language of the later Danish settlers of Jutland. One of the troubles about the word is its very late appearance. It is very common in the 17th century and onwards and at first one was inclined to think that it must be quite a late importation, due perhaps to some settler from the North Country, who brought in a word which rapidly became fashionable and spread all over the Wealden area, but diligent search ultimately discovered a *tenementum atte Gylle*, near the Dicker, in a Court Roll of 1402, and since then one other example has been noted belonging to the year 1485 and one in the 16th century. On the whole it looks as if the name must be native to this area. If that is the case it is probably from a Germanic stem **gulja*, the possibility of whose existence Ekwall suggested in discussing *Gooden* in Lancashire. This is allied to the Swedish *göl*, "pond," MHG *gülle*, "pool." These words Hellquist (*Svensk Etymologisk Ordbok*, *s.v. göl*) takes to be from a *gul*-form parallel to and existing side by side with the more common *gil*- and presumably used with something of the same meaning.

Another lost OE word of which the existence can be proved beyond question is *anger*. There are three examples of *Angram* in the North Riding and Goodall notes three more in the West Riding while Ekwall has one in Lancashire. The early forms of these names are clearly from the dative plural of an OE word *anger*, the same which in its nominative form is found in *Ongar* (Essex). It is also found as the first element in *Ingram* in Northumberland, *Angerton* in Lancashire and in Northumberland. This word is clearly the same as OHG *angar*, ModHG *anger*, "grassland," especially as opposed to forest and to arable land, but also as opposed to swampy or heath-land (*v.* Fischer, *Schwäbisches Wörterbuch*, *s.v.*). More important for our purpose is perhaps the fact that it is common in the forms *Anger* and *Angeren*, *Aengern* (dative plural) in Westphalia and that we have it in the old Latin name *Angrivarii* of a people whose home was on the Weser, and whose name survives in the name of the modern district of *Engern.*

More difficult to distinguish and to interpret is the *reach* of "Heath and *Reach*" in Bedfordshire and *Reach* in Cambridgeshire, with early forms *Reche* and *Rache*, not going back however beyond the 13th century. *Reach* in Bedfordshire is a village lying along a road which runs up a shallow valley. *Reach* in Cambridgeshire stood, as old maps show, on a projecting piece of land, at the very edge of the fens. More important is the fact that it lies at the end of the Devil's Dyke, just where that dyke, having served its purpose, comes to an end. In the Middle Ages the dyke must have been the natural path of approach to Reach which, to this day, is singularly inaccessible. Now we have in the English dialects a word *rake*, "way, narrow path up a cleft or ravine" and a cognate *rack*, from a different ablaut-grade, meaning "narrow path or track." The first is a Scandinavian loan-word from ON *rák*, "stripe, peak," but the second is of such wide distribution in the south and west of England that we must take it to be of native English origin. Corresponding to these two words there may have been in OE words *rǣc*, *ræcc*, *jo*-stems with consequent palatalisation of the *c*, and we may indeed have actual survival of one or other of these in the dialectal words *rache, ratch, reach* used of the "white streak on a horse's face." Such an OE word might well have been applied to the Bedfordshire site, lying along a steep and narrow road up a valley, and to the Cambridgeshire one which could only be reached by the long narrow stretch of the Devil's Dyke.

The last word with which we will deal under this head, and indeed in this lecture, is one which well illustrates the curiously elusive character of many of these place-name elements. In an early Fine (1202) relating to Icklesham in Sussex and in three other ME documents, we have reference to a place called *Cliuex* which, though clearly related in some way to *Cliff* End in that parish, could not be that place. Gradually one came across a number of other examples of the same final element, all from the extreme south-east of the county. There was a *Spelt(h)ex* (1296, 1327) in which it seemed to be associated with OE *spelt*, "corn," a *Ferthex* (*c.* 1270) in which the first element seemed to be OE *fyrhþ*, "woodland," a *Brok(h)ex* (1296) in which it was compounded with OE *brocc*, "badger," or *brōc* in its Sussex sense of "water meadow," a *Brodenexe* (1332) of which the meaning was obvious, and two other cases in which it was almost certainly compounded with a personal name. None of these survives on the present-day map and only one can be located. Mr J. E. Ray has identified *Cliuex* with a field called *Cleeve Axe* marked on the Tithe Award as down in the marshes, not far from Cliff End. This word, as suggested by Ekwall, is no doubt related to the common German word *esch(e)* denoting "corn-field," frequent in Westphalian place-names, found also in OHG *ezzisch* and in the Gothic *atisk*, used in the same sense. What exactly was its meaning in Old English we cannot say. It may have denoted corn-land, it may have meant only pasture-land, but whatever its exact meaning, it is clear that we have yet another of those lost Old English words. This one has the added interest that it was clearly one which was peculiar to a very small band of the early settlers and that, though of such great interest, it has practically disappeared from the face of the land.

These illustrations, ranging over a wide variety of types and touching the life of our forefathers at many different points, would seem amply to demonstrate how rich is the mine of information to be found in the study of our place-names by those anxious to discover the full extent and character of their vocabulary.

LECTURE III

Lines of Interpretation

IN this last lecture one can only touch upon one or two of the problems which face the place-name student and one may perhaps begin with a problem which is perpetually with him, viz. whether this or that element forming the first part of a place-name gives us the name of its owner or is some common word describing something connected with the actual place itself, and that will lead us on inevitably to a second problem, viz. our knowledge of the range and character of OE personal nomenclature.

There are of course many names in which there is only one possible answer to the first question. The personal name may be so obvious or the topography so clear that there can be no difference of opinion. Of the former type it is perhaps superfluous to give an example. Of the latter we may note the case of *Souldrop* (Bedfordshire). The early forms *Sultrop*, *Suldrope*, etc., suggest, since there is no sign of any inflexional vowel between the *l* and the *t* or *d* (*v. infra* 94), that this cannot contain the personal name **Sūla* found in *Sulangraf*, *Sulanford* (BCS 691, 1282), but that we must rather look to such a parallel as *sulhford*

(BCS 166, 1331), which clearly denotes a ford lying in a *sulh, i.e.* a "furrow" or "depression." When we examine the site of *Souldrop*, we find that it lies in a well-marked valley, out of which you make your way by *Souldrop Gap*, and the topographical interpretation of the first element of the name at once becomes clear and certain.

Unfortunately there are a large number of cases in which the solution, in one direction or another, is by no means obvious. When this is the case there is danger of our judgment being biassed by a very natural and human prejudice in favour of the descriptive solution. It is more satisfying to our appetite for the definite, the precise, the picturesque, to think that *Pollicott* in Buckinghamshire means "cottages of the dwellers by the pool," than that it means "cottages of the people of one **Pōl* or **Pōla*," the unknown name of an unknown man. When the latter solution of *Pollicott* was suggested in the *Place-names of Buckinghamshire* the question was quite naturally raised, "Does the geographical situation of *Pollicott* forbid derivation from *pool*?[1]" The answer is that it most definitely does. The *Pollicotts* (in Ashendon) are on a curious steep and isolated hill rising to some 500 ft. We must rest content with the less picturesque solution.

A similar question was raised with regard to *Knotting* (Bedfordshire)[1]. This had been explained as an *ingas*-derivative of a personal name **Cnott(a)*. Could it not be derived

[1] *Englische Studien*, 62, 101.

from OE *cnott* used to mean "hill," cf. North Country *knot*, "hill"? The answer here is a little longer but not less clear. We have no evidence for a word *cnott* in OE meaning "hill." The word is well known but is never used in that sense in the charters or any other Old or Middle English document except *Sir Gawayne and the Green Knight*. The word *knot* denoting a hill is only found in North-western England, in Lancashire, Cumberland, Westmorland and the Craven district of Yorkshire (hence its presence in *Sir Gawayne*), where, as Ekwall has shown (*Scandinavians and Celts*, 40), it is very probably of Scandinavian origin. Elsewhere, not only is it not found in the charters but, so far as the material has been explored, it does not seem to be found in any place-name material at all. It is not therefore in the least likely that we have a word *knot* denoting "hill" in *Knotting* or *Knottingley* (West Riding of Yorkshire), or *Notting Hill* (London), earlier *Knotting* Hill, or in the Berkshire *Cnottingahamm* (BCS 895). On the other hand, we have definite evidence for a personal name *Cnott* or *Cnotta* in *Knottsall* (Worcestershire), *Cnotteshala* in 1294, and in a lost *Knottenhill* in the same county, earlier *Knoteshull*, *Knottenhulle*.

The suggestion of a topographical interpretation of these names was of course in part prompted by the fact that the *Cnottingas*, who gave name to *Notting* Hill, must clearly have lived on a hill (though not a very prominent one). It is true also that *Knotting* is on comparatively high ground and that though *Knottingley* is down by the Aire, practically at sea-level, there are to the west of it, a mile or so away, some small isolated hills, but here we have an example of another of the difficulties which lie in wait for the place-name investigator. Here you have three place-names, presumably containing the same element, which might at a pinch perhaps be represented as having something of the same topography. Does it follow that that element must necessarily describe the topography of these places? Unfortunately not. The similarity in the nomenclature may be a pure coincidence. The odds in these matters are incalculable. A good example of this is to be found in *Strensall* by York. The early forms of this name show that it is identical with *Streonæshalch*, the old Anglian name for Whitby. Now the only other known examples of the first element in these names are to be found in two place-names found in Worcestershire charters, a *streones halh* (KCD 1358) and an entirely independent *streonen halæ* (BCS 1139). It is agreed that **Strēon(a)*, not found independently, is here a pet-form of a rare type of OE personal name, preserved in the compound *Strēonbercht*. Here then we have a pet-form of a rare OE personal name only occurring four times and yet always compounded with *healh*. This cannot mean that it was forbidden to anyone named *Strēona* to possess anything except a *healh*. It is only an extraordinary coincidence.

One other example of such a coincidence may perhaps

be noted. There are in England places called *Ludgershall* in Wiltshire and Buckinghamshire, *Ludgarshall* in Gloucestershire, and *Lurgashall* in Sussex. The early forms of these names all show *Lutegareshale, Lotegareshale*[1], and so does another lost place of the same name in Arkesden in Essex. The first element, whatever it may be, is only found certainly once elsewhere, viz. in the old name *Lutgaresberi* for Montacute in Somersetshire. It is only this last chance survival of an extinct name which saves us from yet another amazing coincidence, for the four times repeated combination of two particular elements, one of which is very rare, is equally strange, whether we interpret the first element as a personal name and say that it means "Lutegar's *healh*," or as a significant word and say it means "grass-land on a *healh* allotted by lot" (OE **hlyte-gærs*)[2].

A good example of the difficulties of calculating the odds as between the personal and place-name solution is to be found in the problem of the name *Goldbridge* in Sussex. There are two examples of this name with ME forms *Goldebregge*, etc. Are we to explain this as the "bridge of *Golda*" or as a compound of *bridge* and some descriptive word? There is a late OE personal name *Golda*

[1] The form *Letegareshale* in Roberts' *PN of Sussex* is an error, while *Litlegareshale* (DB) for the Wiltshire name is almost certainly corrupt.

[2] Wallenberg (ZONF 4, 287) has shown one of the difficulties involved in this latter solution. Equally important is the fact that we always have *gares*, never *gers* or *gars*.

and it may be that two persons so named owned or built bridges in Sussex. On the other hand, there is some evidence, though not very definite, that there was an OE **golde* used of some yellow-flowering water-plant. Were the bridges so called because by them there chanced to grow a plant of this kind and in such profusion as to suggest the name? It would be impossible to calculate the odds for or against either solution, even if one had been through that course at Monte Carlo which Dr Chambers has recently suggested would be a useful part of the scholar's equipment.

But those concerned with these problems are faced with other difficulties beside those of calculating probabilities. There are more subtle snares, even for the most wary. A year or two ago Zachrisson noted some interesting examples of river-names containing fish-names and went on, naturally enough, to say "we may compare such piscatorial names as *Graylingwell* (Sussex)[1]." Ritter, two or three years previously, in a review of Roberts' *Place-names of Sussex*, had suggested that here we had the genitive plural of a patronymic **Grǣgelingas* from a lost OE personal name **Grǣgel*, connected with the adjective *grǣg*, "grey," and cognate with OGer *Grawilo*, hence "little grey one." At first sight one is bound to fall a victim to the interest of Zachrisson's solution. Unbeknown to either of them two pieces of evidence were lying in wait which vindicate the less picturesque solution. In the first place *Graylingwell*

[1] ZONF 2, 146.

is not on a stream (*well* can of course mean "stream" in OE) but is the name of a spring, as the 6″ O.S. map shows, and there can be no *grayling* there. In the second, there is a place elsewhere in the county, now called *Grainingfold*, with early forms *Greling(e)fold* and the like, exactly of the same type as those for *Graylingwell*. Here also there cannot have been any *grayling*, but there may well have been a man *Grǣgel*, who gave name to the place either through himself or his descendants. Some other man also so named was similarly responsible for *Graylingwell*.

It is well to remember that in some cases no definite answer is or ever will be possible. *Oving* in Buckinghamshire is a case in point. The early forms of this name show that in form at least it is identical with *Oving* in Sussex, both going back to OE * *Ufingas* and both being pronounced at the present time *Ooving*. In the *Place-names of Buckinghamshire* the former of these names was derived from a personal name *Ufa*, while in the *Place-names of Sussex* the same etymology is suggested for *Oving* in that county, with the additional note that *Ufa*, or rather *Uf(e)*, the same name in its strong form, which would equally well explain *Oving*, is actually found in a ford called *Ufesford* in an OE charter (BCS 50), just a mile or two away from *Oving*. Zachrisson, in a recent article on "Early Teutonic Tribal Names[1]," in which he shows how many of them are of geonymic origin, has suggested that as the Buckingham-

[1] *Studier tilegnede til Axel Kock*, 490 ff.

shire *Oving* stands on a hill rising to 500 ft. the name *Ufingas* is in this case to be associated with the OE adverb *ufan*, "above," and the name interpreted in the same way as *Epping* (Essex), viz. "dwellers on the high ground." That may well be so, but who could say that it is so with certainty? The utmost that we really can say is that *Oving* in Sussex, down on flat land, *must* mean "people of *Ufa*," while *Oving* in Buckinghamshire *may* mean "people of *Ufa*" or it *may* mean "dwellers on the hill."

One holds no brief for personal as against descriptive elements and when one is faced with a difficult name one may at times be too ready to seek a way out with the aid of an unrecorded personal name, but it is clear that there is an equal danger, even for the most prudent, in resorting too readily to a geonymic alternative. *Saunderton* in Buckinghamshire with early forms *Santresdon, Sandresdon, Sandreston* presents difficulties, but could be explained if we assumed that the name was comparatively late in origin and contained the continental Germanic name *Sandhere*, the name of some late immigrant from the continent. That may or may not be right. The alternative explanation which has been suggested, viz. that it is from OE *sand-hrīs-dūn*, "sand(y)-brushwood-hill[1]," cannot hold good, for Saunderton is on the chalk and local enquiry shows that there is no trace of those patches of sand which one sometimes gets on top of the chalk.

[1] ZONF 4, 93.

Sometimes it is not a question of topographic versus personal-name explanation but of difficulties arising from too great readiness to resort to this or that topographic explanation even of what is clearly a significant place-name element. This is well illustrated by the history of the interpretation of *Fyning* (Sussex). Ekwall, in his *Place-names in -ing*, very tentatively explained this as an *ingas*-derivative of a personal name *Finn*[1]. Zachrisson suggested that it was really a derivative of *fen*, "marsh[2]," and quoted in support of this fact that in the neighbourhood are Trotton Marsh and Milland Marsh, but the first is a mile and a half to the north-west on the other side of a hill 450 ft. high and the second is still another two miles to the north-west, so that they give no help in the matter. At the same time Zachrisson did good service in showing that *Fyning* must be a significant word, for in the Subsidy Roll of 1296 we have a Robert *atte Finnyng* in Burton or Bignor. After the difficulty of the "fen" view was pointed out, Zachrisson abandoned it in favour of another geonymic derivative of *finn*, denoting "hill, hilly woodland," and in the paper on "Early Teutonic Tribal Names" just referred to, *Fyning* is quoted as giving support to the interpretation of *Phinnoi* (Ptolemy) and *Fenni* (Tacitus) as "those who live on woodland or high ground[3]." But what is the topography of the places

[1] *PN in -ing*, 63.

[2] *English PN in *Visk, *Vask*, 53–4.

[3] *Op. cit.* 494.

bearing this name in Sussex, for there are several of them of which only two were known to Zachrisson? *Fyning* itself is low down on the slope of a gradually rising hill. *Vining* Rough (earlier *Finnings*) in Easebourne is on high heath-land. There was also a lost *Vining* in Westhampnett mentioned in a *via inter Vining et Schepewyk. Shopwyke* is to the south of Westhampnett in flat land and there can be no question of any hill here. Even more definite is a field called *the Vinnings* in the Sidlesham Tithe Award. Sidlesham parish is in the heart of the flat Selsey peninsula. These names can have nothing to do with hills. The word is rather to be connected with OE *fīn*, "wood-heap," hence "place of wood-heaps," a woodland term or, if the vowel was originally short, and the evidence is rather in favour of that, then Ekwall suggests in view of the new evidence that it may contain a plant-name cognate with Norw *finn*, "nardus stricta," MDu *vinne*, "bristle of an ear of corn," hence "place where coarse grass grows." No certainty is possible except that neither the *fen* nor the *hill*-generalisation will fit the case.

With these difficulties before us, is it possible to suggest any helpful line of approach towards the great problem of distinguishing personal and place-name elements? Are there, for example, certain second elements which require a personal name before them and others which require some significant word? Diligent search has yielded only two elements in which it may be safe to lay down anything like a general principle.

stow, "place," does not seem to be compounded with a personal name, unless it happens to be a saint's name, the land being dedicated to his or her service, as in *Hibbaldstow* and *Virginstow*, the first containing the English saint's name *Hygebeald*. Rather, it is compounded with such elements as *cot* in *Costow*, *wīc*, "dwelling, farm," in *Wistow*, *mōt*, "meeting," in *Moustow*, or a tree-name as in *ellen-stow*, "alder-tree place," first suggested by Zachrisson for the difficult *Elstow*, earlier *Elnestowe*. Even here however there are exceptions. *Bunsty* in Buckinghamshire can hardly be explained without recourse to OE *Būna*, which was certainly not a saint's name. If the place were in the south of the county, one might perhaps explain the first element as OE *bȳne*, "cultivated," but up on the borders of Northamptonshire such a development of OE *bȳne* is impossible. So also Alstoe (Rutland), Broxtow (Notts) contain respectively OE *Aelfnōþ* and *Brocwulf*. It is at least suggestive that all three of these are hundred-names, so that the personal names may have been those of a sometime "hundred's-man."

The suffix *stede*, "place, site," offers similar possibilities for most parts of England. It is usually compounded with words like *hām*, "home," *w(e)orð*, "enclosure," as in *Hampstead* and *Worstead*, or with some tree or plant-name as in *Ashstead* or *Hempstead* in Kent which, as Wallenberg has recently shown, contains in some instances the word "hemp[1]," but here one has to make the important exception, that in East Anglia, Kent, Sussex, and Surrey there is a sprinkling of *stead*-names containing a personal name, such as *Bersted* in Sussex, from **Beorga* (*v. infra* 100). The explanation of this area of exceptions lies almost certainly in the fact that these were the counties first settled and that the first settlers used the term of an "estate" in a way which soon became obsolete.

There is another more helpful line of approach to the problem, at least in certain cases. If we have a personal name in Old English it must have had a genitive singular ending in either *an* or *es* or *e* and it is dangerous to postulate a personal name for a place-name preserved only in ME forms unless it shows signs of that inflexional syllable. The *es* is easy enough to recognise as *es* or *s*. The *an* usually becomes *en*, more often simple *e*. If we do get, fairly persistently, an *e* between the two elements of a place-name in its ME forms we are probably justified in presuming that in OE we had either a genitive plural in *a* of some common or proper noun, or an inflected form in *an* or in *e* which was the genitive singular of a personal name. We could argue back from the persistent *e* of the ME forms of *Shellingford* (Berkshire), *Sceringeford* and the like, to OE *scaringaford* with genitive plural in *a*, even if we did not know that there was such a form in an OE charter

[1] *Studia Neophilologica* 1, 35 n.1.

(BCS 683), and we could argue back from the persistence of forms like *Civelei* for *Chieveley* in the same county, to OE *Cifanleage*, without being aware of that form in another OE charter (BCS 1055). Similarly, we can argue back from *Olvelei* and *Wulvele* for *Woolley* in Berkshire to *wulfaleage* (BCS 762), "wolves' clearing," or from the persistence of medial *e* in such forms as Domesday *Baltredelege* for *Balterley* (Staffordshire) to the genitive feminine singular of OE *Bealdðrȳð* in *Baltryðeleage* (KCD 1298).

The significance of this medial *e* in such names as these has sometimes been doubted (cf. *supra* 4–5), but without sufficient justification. To put the matter to the test, let us take for example all the place-names in Worcestershire for which we have OE forms, of whatever date or authority, and there are a good many of them, containing an inflexional *a*, *e* or *an*. We shall find only two examples, viz. *Canterton* from OE *Cantwaratun*, and *Redmarley* from OE *Reodemæreleage*, which do not fairly persistently show *e* in their ME forms, *i.e.* two out of some forty examples. On the other hand, where we have no inflexional *a* or *e* or *an* in OE, neither do we as a rule have an *e* in ME. Out of some forty examples the only exceptions that have been noticed are *Elmley*, *Shrawley*, *Swinford* and *Lindworth* in which we have compounds of OE *elm*, *scræf*, *swīn*, and *lind*. The charters show *scræfleah* and the like, the ME forms show *Shrewelege*, etc. fairly frequently. The hesitation may well be due to the existence even in OE of alternative forms such as *swīn-ford* and *swīna-ford,* "pig-ford" and "pigs'-ford." (Already in Heming we have *Elmelege.*) An inorganic *e* also appears in *Witley* and in *Wychbold.* For *Ockeridge* and *Cromer,* going back to OE *heafuchrycg* and *cronmere,* we have in each case only one ME form and that of the 13th century.

It is clear then that ME medial *e* cannot be lightly set aside, and this is true even when we have a fairly even division between ME forms with and without *e. Waldridge* in Buckinghamshire appears as *Waldruge* in Domesday, then has one form with *e* in the 12th century, and five with *e* and four without *e* in the 13th century. We should not be justified in saying that this must be a compound of OE *weald,* "forest," and *hrycg,* even though we did not know that the place was called *wealdanhrycg* in an original 10th-century charter (BCS 603). It is the knowledge of this principle which would make one suggest a different derivation for *Ripton* (Huntingdonshire) from that for *Ripley* (Derbyshire), even if one did not know that the latter goes back to OE *Rippanleah* (BCS 1361) as against *Riptun* for the former, for *Ripley* is persistently *Rippelye* and the like in ME, while *Ripton,* with one exception, shows no medial *e* in its forms.

Middle English scribes did many wanton things and it would be absurd to try to defend every medial *e* in the ME forms of place-names and say that it must be significant, but it is clear that in the matter of medial inorganic *e* the scribes were not quite so wanton as they have sometimes been thought, or, to put the matter in another way, medial *e* in ME place-name forms is always worthy of very careful consideration.

Before leaving this question of medial inflexional syllables, a word must be said of a type of place-name which has to be borne in mind in balancing the probabilities of personal as against descriptive elements in the first part of a place-name. Some few years ago Ritter[1] demonstrated the presence in the terminology of the OE charters of a large number of genitival compounds, *i.e.* compounds in which the first or defining element in a toponymic phrase is in the genitive case. He gave numerous examples of such expressions as *broccholes weg*, "the road of the badger-hole," or *Fitelan slædes crundel*, "the chalk-pit of Fitela's valley." Unfortunately none of his phrases are quite clear examples of place-names pure and simple. In setting forth the boundaries of a piece of land it may have been useful to refer to a particular road as the "road of the badger-hole" or to some particular chalk-pit as "the chalk-pit of Fitela's valley"; it does not necessarily follow that the name for that road in local usage was "Brockhole's-way" or the name for the pit "Fitela's-slade-chalk-pit." They may be what Redin has rather happily called "indications made for the occasion." All the examples given, and there are many of them, contain three elements, and triple element place-names are very rare indeed. But though the evidence

[1] *Vermischte Beiträge*, 155–6.

thus adduced does not perhaps quite prove the case, there can be no doubt that genitival compounds did exist in Old English and that we have got to allow for them in our place-name interpretations.

No reasonable explanation of *Beaconsfield* (Buckinghamshire) can be offered except to suggest that it is the "open-land of the beacon." It is a place which stands high and might well have been used for some form of signalling by fires. It is very difficult not to think that *Beeston* (Bedfordshire) is from OE *byges-tūn*, "farm of the bend," referring to the bend in the Ivel at this point. We may note also the existence of a slightly different type of genitival compound in the old names *Andredesweald* and *Andredesceaster* for the Weald and for Pevensey. These are obviously genitival derivatives of *Andred*, the name for the forest which stretched from Kent to Hampshire in Anglo-Saxon times.

Two further examples may be added. For *Maplesden* in Ticehurst (Sussex) we have early forms like *Mapeltresdenne* which must go back to OE *mapuldresdenn* or *mapeltrēo(we)s-denne* and be interpreted as "mapletree's swine-pasture." *Frithsden* in Berkhampsted (Hertfordshire) is *Frithesden* in 1293 and in the same entry in the Charter Rolls we have reference to a wood called *le Fryth*. It would seem to be practically certain that *Frithsden* is the "valley of the woodland (OE *fyrhþ*)."

It is clear therefore that a genitival form in the first

element of a place-name does not necessarily imply a personal name. We have ceased to use inflexional genitives in Modern English in the case of words denoting objects without life, and it is repugnant to our speech-instinct to think that people could ever have spoken of the "maple-tree's *denn*," but our ancestors felt no such inhibition and could use such inflexional forms, though they clearly preferred to use on most occasions the uninflected form "mapletree-*denn*."

We must now pass to another important aspect of the matter in hand. A good many of our difficulties arise from our imperfect knowledge of the full extent of Old English personal-nomenclature, and objection has from time to time been taken when a difficult place-name or series of place-names have been explained by reference to an unrecorded personal name and especially to unrecorded personal names of a rare or exceptional type.

In answer to this one must first make the general observation that just as our last lecture showed us that we could prove the existence in the speech of our forefathers of many hitherto unknown or unsuspected words or even, as in the case of the *-et* suffix, discover a new method of forming a whole series of noun-derivatives, so we are bound to find large gaps, not only in our knowledge of individual personal names, but even of personal-name formations during the same period. It is important to remember that we have as yet no complete list of personal names found even in Old English documents. Searle's *Onomasticon* for example does not give *Cidda*, found in the calendar of St Willibrord (*c.* 710), *Byrngȳð*, the name of the nun addressed by Aldhelm in the preface to the *De Laudibus Virginitatis*, *Wyrmhere*, mentioned in *Widsith*, or *Beald*, found in the Leechdoms. It includes none of the names of the 11th-century burgesses of Colchester set forth in the DB for Essex. The last three personal names might have been expected, but when *Cidd(a)* was required for *Chidlow* (Cheshire), earlier *Chiddelowe*, or for *Chidgley* (Somerset), earlier *Chideslegh*, he would have been thought a rash man who postulated it. Similarly, the Colchester burgess-list gives us the names *Blanc* and *Berda* required for *Blankney* (Lincs), *Bardney* (Lincs) and *Barnstaple* (So).

One might suspect the name *Beorga* required for *Bersted* in Sussex, with a form *Beorganstede* in a 10th-century copy of a 7th-century charter (BCS 50), for personal names with the element *beorg*, fairly common on the continent, are unknown in England. There is however one interesting piece of confirmatory evidence. The name of the princess commonly known as *Ethelburga*, the daughter of Aethelbert of Kent, who married the Northumbrian king Eadwine, is persistently spelled *Aedilberga* in the oldest manuscripts of Bede, showing this very element. Later scribes, not familiar with this name-formation, which must already have been archaic, replaced it by the common *burg*.

Two other examples of our ignorance of the elements

used in OE compound names may be noted. There is a place *Lundsford* in Sussex with early forms *Lundredisford*, etc., which must contain an OE personal name **Lundrǣd*. No other compounds of *Lund-* (cf. ON *lundr*, "mind, disposition") are known in English and we have to seek parallels in the Frankish *Lundoaldus* and the Swedish *Lundbiǫrn*. Similarly, we have in a post-Conquest English document the name *Ristegar*[1], clearly an Anglo-Norman spelling for *Rihtgar*, our only evidence for *Riht*-names in English.

Further evidence of the gaps in our knowledge of OE personal names is to be found in the number of cases in which a personal name required for the explanation of place-names has been found on record in the early ME period, usually in the 12th century, evidently a survival from pre-Conquest days. A name *Blic* was assumed for *Brickley* (Worcestershire) and, in a diminutive form, for *Blickling* (Norfolk) before record was found in the 12th century of land in Felstead (Essex) called *terra Blik, i.e.* "Blik's land." The early forms of *Wakeham* (Devon and Sussex), *Wakefield* (West Riding) and *Wakeley* (Hertfordshire) would lead one to infer an OE personal name *Waca* even without the evidence of a name *Wache* found in a 12th-century Norfolk charter to confirm it. There is a series of names—*Poppinghole* and a lost *Pophall* in Sussex, *Popham* in Hampshire and *poppinghangra* (BCS 923), probably to be identified with *Popes* Wood in Berkshire—

[1] Harl. 742, f. 287 b.

whose early forms suggest an OE personal name **Poppa*, of which we have a diminutive **Popela* in *Poppleton* (Yorkshire). In Essex we find *Poppe* used as a personal name in the 12th century. *Pounsley* in Framfield (Sussex), *Powtenstone* in Groveley (Wiltshire), earlier *puntes stan* (BCS 935), and *Poyntington* (Somersetshire) all require a personal name *Punt* for their explanation, as does *Punteles treowe* (BCS 787) in Hampshire, where we have a diminutive of *Punt*. There was a man called *Punt* living at Andover *c.* 1106. For *Mackney* (Berkshire), *Maccanig* in BCS 810, we should postulate a personal name *Macca*, but this is on record in early ME as *Macke*. So for the series *Natsworthy*, *Natson*, *Nottiston* in Devon, with early forms *Notesurde* and the like, we need a personal name *Nott*. Such is to be found in *Godwin filius Nhott* in a post-Conquest Berkshire document, which at the same time indicates the etymology of the name. The *nh* is for *hn*, the name coming from OE *hnot*, "bald-headed." The Devonshire *Noggacote* looks hopeless. The DB form is *Nochecota* with later forms *Noggecote*, *Nuggecote*. One could only solve it by postulating a pet-form *Nogga* from such an OE name as the feminine name *Nōþgȳþ*. The name *Nogga* is on record as a woman's name in use after the Conquest. The simplex for the diminutive *Grǣgel* postulated above for *Graylingwell* and *Grainingfold* is unknown in Old English but soon after the Conquest we get *Sewold Graiessune* and, somewhat later, *Graiue*, *i.e.* *Grǣg-giefu*, a woman's name. *Throckley* (Northumberland) and *Throcking* (Hertfordshire) seem to require a personal name *Throcc(a)*. The early forms for the first are *Trocchelai* (1160), *Trokelawa* (1176) and it is on a hill as its name indicates. *Throcking* (Hertfordshire), DB *Trochinge*, is at the highest point in the neighbourhood. Neither of these can contain the word *þrocc*, "drain," noted in our last lecture. On the other hand, there is a personal name *Troke* on record in the 12th century. The early forms of *Gransden* (Huntingdonshire) suggest a personal name *Grante*. There was a person of that name, mentioned in the Croyland cartulary, living not far off in the 12th century, and someone of that name was responsible for the 13th-century Worcestershire field-name *Grantesforlong*. Mr *Trump* of *Trumpington* is not the invention of the personal-name enthusiast. There was a man of that name who witnessed a charter in the Thorney cartulary and who, with the usual confusion of *p* and *w*, appears in a late copy of that charter (BCS 1297) as *Trump*. The personal names *Uuordgine* (11th cent. Bury Survey[1]), *Wrtheue* (1222 DB of St Paul's) must go back to OE *Weorðgiefu* (cf. continental *Werdhilt*, *Vertleuba*, *Verdthun*, Förstemann, *PN* 1559–60) and gives an English example of a personal-name element *Weorð* which provides the best explanation of Worthing (Sx).

Another way in which these personal names tend to reveal and vindicate themselves is to be found in the

[1] Ex inf. Mr D. C. Douglas.

formation of small groups in which a particular name and its derivatives are found concentrated in some limited area. Stenton gives examples in his chapter on the English Element (IPN 44–5), and Zachrisson has noted the prevalence of personal names containing the element *Gef-* in the neighbourhood of Eastbourne. One or two further examples of groups containing personal names not otherwise on record may be noted. Ekwall pointed out that *Wartling* and *Worsham* (OE *Wyrtlesham*) close by both contained the same unknown personal name * *Wyrtel*. This name is found once again in *Ridlington* in Ambersham at the other end of the county. For *Whatlington* in Sussex one had to assume an OE personal name **Hwætel*, a derivative of *hwæt*, "active." The probability of such a name was very definitely increased when one found in the next parish a place called *Waddles Wish* whose early forms suggested the same personal name. We might have doubts about a personal name **Climp*(*e*), the head of the clan of *Climping* in Sussex, were it not for the clear presence of such a name in *Clemsfold* in the same county going back to *Climpesfaude*, and we might doubt the existence of a *Pefen* in *Pevensey* were it not the case that we also have a *Pensfold* going back to earlier *Pevenesfald*.

The knowledge of the existence of such groups can be of help at times in solving a problem. *Fitzhall* in Iping (Sussex) requires a personal name **Fit*(*t*) to explain its early forms *Fitteshale*, etc., from 1279 onwards, and so also does *Fitzlea* in Lodsworth in the same county. This gives us the otherwise unknown simplex of the rare *Fitela* found in *Fittleworth*. One hesitated in the case of *Polehanger* (Bedfordshire), now pronounced *Pullinger*, between a personal name *Pōla* or *Pulla* and the significant word *pōl*, "pool," though it was hard to find topographical justification for the latter. It is perhaps worth noting that four or five miles away there is a place called *Pulloxhill* which clearly contains a personal name *Pulloc* (the topography forbids *pulloc*, "little pool"), the same name which is found in *Poulston* in Halwell (Devon), DB *Polocheston*, again with no possibility of a "pool." The Devon name, containing *Pulloc* or possibly *Polloc*, goes in its turn with the *Pulworthys* in Hatherleigh, earlier *Poleworthy*, which are both on hills, and with *Polsloe* in Heavitree, DB *Polslawe*, of which the site is ambiguous, but in which the genitival form makes a descriptive element unlikely.

The existence of most of these names would now be generally accepted. More difficult is the problem of one or two name-types whose existence seems to be required for the explanation of certain difficult place-names and probably also for certain personal names actually on independent record. It is generally admitted that in Teutonic personal nomenclature you might form a new name from an earlier one by what may be called an *n*-extension of the stem. Thus from a name *Pippa* you could form a new name *Pippen(e)* and from *Tadda* a derivative *Tadden(e)*, the new forms possibly being regarded as pet-forms of the others. For the explanation of certain personal names, not only in English but also in the other Germanic dialects, similar *r*- and *s*-extensions have frequently been postulated by scholars both English and foreign, though with some hesitation, especially in the case of the *r*-suffix. Recently Zachrisson has denied altogether the existence of these *r*-[1] and *s*-extensions in English[2] and in the Germanic dialects generally, and has with much ingenuity endeavoured to explain away all possible examples of their occurrence.

It is therefore worth while examining afresh the position with regard to the more important of these, viz. the *r*-extensions. Zachrisson's view is (i) that these *r*-extensions are not found in any of the other Germanic dialects, (ii) that there are no clear cases of such names on independent record in English, and (iii) that all English place-names for which such personal names have been adduced can be explained in other ways.

Fortunately it is not necessary to enter into the difficult questions involved in the first of these statements. Even if *r*-extensions were proved beyond question never to have been used in the other Germanic dialects, that would not prove that they did not exist in English. That this is so can be seen very clearly from another series of personal names. We have a series of well-established OE personal names *Monnede, Luhhede, Lullede, Ucede* and probably also

[1] ZONF 4, 245 ff. [2] *Festskrift til Finnur Jónsson*, 316 ff.

Wipped, on independent record, which are clearly extensions of OE personal names *Monn*, *Luhha*, *Lull*, *Ucca* and *Wippa*, also on record, beside a considerable number of others which are found only in place-name compounds. This *-ed*(*e*) suffix has no parallel in the other Germanic dialects and yet its existence stands beyond question. If we could form extensions of this kind within English itself, why not similar extensions in *or* or *ra*, especially as there were a good many common words in use in the language in which such a stem-extension had taken place without any obvious distinction of meaning between the new and the old words? Side by side with *sige* we have *sigor*, side by side with *telga* we have *telgor* and *telgra* and, as Zachrisson himself has recently suggested, side by side with *pōl*, "pool," one probably had an OE **pōlra*, a stem-extension which would explain the word *polre* or *polder* which was discussed in our last lecture[1].

We may pass on then to the second objection. We have in two early charters, as transcribed by Bond in the British Museum facsimile edition, personal names *Hymoran* (dat. sing.) and *Dilra*. Unfortunately as the charters now stand it is impossible to be certain of more than *H..mo..* in the one case and *Di..ra*[2] in the other, with the knowledge that the missing letter in the second case is a

[1] *Early Teutonic Tribal Names*, 493 and *supra* 51.

[2] If the name is *Diera* it cannot, as has been suggested, be the name of the same man who signs BCS 226 and 241, for the latter was a bishop and this man was not.

tall one which must be either an *l* or a *t* or an elongated *e*. Bond's reputation as a palaeographer stands so high and he is so scrupulously careful to leave blanks where he cannot decipher the letters that it is very unsafe to question a reading which he printed without any hint of uncertainty[1]. There is the third case of the name *Tepra*. This is found twice, once in an 11th-century fabrication and the second time in a copy of an 8th-century charter preserved in Heming (*c.* 1100), (BCS 32, 157), but Zachrisson notes that *p* and *r* are much alike in the handwriting of the first of these charters, and suggests that *Tepra* is the copyist's blunder for *Teppa*, and dismisses the second signature as a similar blunder which has no authority as it occurs in a Middle English compilation.

We may deal first with the second signature. It is found in a text of *c.* 1100. Judged both by diplomatic and linguistic tests, the genuineness of this charter is beyond question and the scribe has preserved with more than usual fidelity the extremely archaic orthography of his original. The whole character of the document and the conditions under which it has been transmitted forbid the assumption that *Tepra* is a blunder for anything else. The first charter in the form in which we have it is a fabrica-

[1] The *Hymoran* charter has been treated with gall at some stage in its history in order to bring out the reading at this point. Possibly it was used by Bond himself to bring out the letters which he then recorded.

tion but its witnesses include names derived from a lost charter of Aethelwald of Mercia. It constitutes therefore independent evidence of a personal name *Tepra,* if such evidence were needed.

With regard to Zachrisson's statement about the handwriting of this latter charter, it is true that when the curved second stroke of the *r* is carried round a little farther towards the left than is necessary, it is on its way towards becoming a *p*, but even when you note this, there is really no case in the charter where you could not readily distinguish *p* and *r*. Far more important however is the fact that no attempt is made to link on a *p* to the following letter, whereas with the *r* you always get a third stroke added carrying it on towards the next letter which follows. You really cannot mistake a *p* for an *r* in this handwriting. It is equally important to remember that the 8th-century charter from which the witnesses named must ultimately have been derived would be written either in uncial or in an early form of insular script in which *p* and *r* are quite distinct, and the copyist would not be in the least likely to confuse them. With regard to the second point, all our Old English cartularies are Middle English compilations, for they were all transcribed after 1100, but that does not mean that they are necessarily unreliable. The Worcester cartulary is second only to the *Textus Roffensis* in the quality of its reproduction of the charters which it incorporates. Further, if *Tepra* is a mistake you have to believe that two independent copyists each replaced a known name by an unknown one, thus defying the old rule *præstat lectio difficilior*. There is no reason to reject *Tepra*, and it can readily be explained as an *r*-extension of an OE personal name **Tæppa*, **Teppa*, which, though not on independent record, is generally recognised.

The third line of attack can best be illustrated by the treatment of the place-names *Peppering* in Burpham (Sussex) and *Peperharrow* in Surrey, with early forms *Piperinges* (in a late copy of an 8th-century charter) and *Pipereherghe* (Domesday).[1] Four possible solutions of the first element in these names are offered, (i) that it is a lost hill-name, (ii) that it is a lost river-name, (iii) that it is the common word *pipere*, "piper," (iv) that it is that word used as a personal name. The first two alternatives would seem to cancel one another out. *Peperharrow* is on a hill. *Peppering* is on the Downs above the Arun, of which we know the earlier name to have been *Tarrant*. If it takes its name from the locality, it must be from the hill on which it stands and not from the river below. There is however a second

[1] *Pepperscombe* in Steyning, earlier *Piperescombe* (1425), quoted by Ekwall (*PN in -ing* 61), need not be considered. It takes its name from the family of one John *Pyper* who held a messuage here in 1354. *Pipernæsse* (KCD 737), found in a Kentish charter of the year 1023, affords no sure foothold for argument. We do not know its site, and the text of the charter is ME with a few old forms preserved in it. All we can say is that it is somewhere in the flat land between Sandwich and the sea. There can have been no hill.

Peppering in the east of the county with similar early forms, not known to Ekwall or to Zachrisson. This is *Peppering Eye* (with Lower *Peppering Eye*) in Battle, down by a stream. The stream is called *aqua de Piperinghe* (12th), *water of Peperenges* (12th), and the place is called *Piperinge* (12th), *Pipringey* in 1383. *Piperinges* may actually have been the name of the stream itself but we know of no stream-names of that form. More probably the river, and the "ey" or marsh land by it, were so called from a place *Piperinges*. In any case if the name has any geonymic significance it can only have to do with a stream. It is clearly unreasonable to think that *Piper* is a lost name of both a hill and a stream.

As parallels for *pipere*, "piper," in place-names Zachrisson quotes an interesting series of place-names from Old English charters of the type of *tannera hol*, "tanners' hole," *bȳmera cumb*, "trumpeters' valley," containing the genitive plural of an Old English occupational name. It is very difficult to see however how these can be used in explanation of the names in question. *Peperharrow* contains as its second element the word *hearg* discussed in our last lecture. It is an ancient heathen site and it seems quite impossible to believe that such a site could have been known as "Pipers' heathen fane or grove." Neither does it seem possible on this basis to suggest a satisfactory interpretation of *Pippering*. This is a very old name. One can hardly believe that when it was first used one could speak of a group of settlers as "the people who gather round the piper(s)," for that is the only meaning that could be given to it.

The fourth suggestion, that the common word "piper" (OE *pipere*) is here used as a personal name, is contradicted by the archaic character of the names *Peperharrow* and *Peppering*, and in the case of *Peperharrow*, which shows no sign of a genitival *s*, it is contradicted by the form of the name. Further, one must not quote in support of derivation from a personal name *Pipere* the 11th-century *Aþelward Stamere*, for that is a nickname of a type which only came in after the Danish settlements. There is no other authority, early or late, for personal names of the agent type in the OE period.

But perhaps the most conclusive is the existence, also in Western Sussex, of a place called *Lippering*, not recorded in Ekwall or noticed by Zachrisson. This goes back to 12th-century *Liperinges*. It is only reasonable to look upon this as a place-name of the same type as *Peppering*. Now *Lippering* is neither on a stream nor on a hill. It is half a mile south-west of Birdham on dead level ground. One cannot suggest that it is a derivative of OE *hlēapere*, "runner," for the forms forbid that. The only reasonable explanation of this name and of the two *Pepperings* and of certain other names of the same type would seem to be that they contain personal names *Pippra* and *Lippra*, or possibly *Pippor* and *Lippor*, which are *r*-extensions of the personal names *Pippa* and *Lippa* found in *Pippanleah* (BCS 1125), in *Pipingminstre* (BCS 729), now *Pitminster* (Somerset), in *lippan dic* (BCS 924) and *lippanhamm* (BCS 629). A man bearing such a name may well have been the owner of the *hearg* at *Peperharrow*, the counter-part of *Pæccel* who owned the heathen site of *Patchway* in Falmer (*v. supra* 59).

In thus defending the existence of *r*-extensions in OE personal names one does not want to be misunderstood. They are clearly very ancient (hence the fewness of the examples on independent record) and should not be lightly used in explanation of difficult names. Zachrisson may well be right in his scepticism about some of the place-names for which a personal name of this type has been or might be postulated, and he offers explanations of some of these to which no objection can be taken. All one is anxious to do is to make it clear that one must still allow the existence of *r*-extensions of OE personal names, even on this evidence, quite apart from anything else, and not try to explain them away by alternatives which raise more difficulties than they seem at first sight to solve.

We may now pass on to another problem which is of perennial interest, viz. the interpretation of the element *ing* in the various positions and forms in which it appears in place-names. Two important books have been devoted to this subject alone, Ekwall's *Place-names in -ing*, now well established as laying the necessary foundations for all study of place-names with final *-ing*, and, more recently, Sigurd Karlström's full and careful discussion of *Old*

English Compound Place-names in -ing, admirably documented and wisely cautious. All that one can attempt to do in this lecture is to try to throw a little light on one or two aspects of the problem from evidence which has come recently under one's notice.

A few years ago in the Introduction to the *Place-names of Northumberland and Durham* the present writer suggested that one could only explain such names (found in OE charters) as *cyneburgingtun*, *werburgingwic* (formed from women's names) and *bisceopingdene* (probably formed from the common word *bisceop*), with the suffix *-ing*, on the supposition that *ing* pure and simple, as distinct from *inga*, had no patronymic force but was really genitival in force, so that these names meant "enclosure and farm of Cyneburh and Werburh," "bishop's *denn* or *denu*," and endeavoured to clinch the matter by reference to the 7th-century charter (BCS 97) in which a grant is made at *Wieghelmestun* and the charter is endorsed, in a hand of the 10th or early 11th century, giving the name of the land as *nunc wigelmignctun* (*sic*), in which *ing* has taken the place of earlier *-es*. Owing to textual difficulties in the charter pointed out by Karlström[1] it is not certain that this charter can be thus used, but other evidence is not wanting. Birch no. 449 is a grant by Aethelwulf to *Badonoð* his apparitor of a villa and certain acres near Canterbury. The charter is endorsed *Badenoðingland* in a contemporary hand (*v. B.M. Facs.* II, xxix).

[1] *Op. cit.* 60.

Karlström has noted *plumwearding pearrocas* (BCS 346), which looks like "plum-warden's," *i.e.* probably "gardener's," enclosures, where we certainly cannot have a patronymic, and the probable case of *Cuðrincgdun* in one charter corresponding to *Cuðricesdun* in another (BCS 378, 496)[1], to which we may probably add *Offerton* (Worcestershire), which appears in pre-Conquest days as *Alðryðetune* but is already *Alcrinton* in Domesday, and *Cudley* in the same county, where it seems that the Domesday and following forms, *Cudelei*, etc., can hardly be merely a reduction of *Cudinclea* (BCS 1298), but must rather be taken as from an alternative form *Cudanleah* with the usual genitival *an*.

In working recently on the place-names of Sussex one has come across three examples in Middle English documents of similar formations in which *ing* seems clearly to have nothing more than some rather indefinite linking force. In a charter belonging to Sele Priory of date *c.* 1250, we have mention of a *Heregrauingfeld* which must have been close by the place called *Heregraue* in a fine of 1231, the *heregraf* of the bounds of the Washington charter (BCS 834). In the Chichester register known as Liber E (14th century) we have mention of *Wychamyngbrok* in connexion with the place now known as *Wyckham* in Steyning. In an inquisition of 1314 we have mention of *Rotforthyngwode*, now *Raffling* Wood in Petworth, which is close by *Ratford* Farm,

[1] *Op. cit.* 146, 168.

earlier *Rotford.* Here an element *inga* (gen. pl. in its form) would seem to be out of the question and the connecting element must be *ing* "belonging to, having to do with," or the like. Farms would have their woodland, their open land, their water-meadows, and it would seem that in referring to such, instead of calling the water-meadows of Wyckham, *Wychambrok,* you might quite naturally refer to them as *Wychaming-brok.* Whether these formations were really alternative to actual genitival compounds such as *Wychamesbrok* it is impossible to say with certainty, but it does not seem very likely. These Middle English examples make me doubt whether it was wise to suggest that the *ing* in such a phrase as *Werburgingwic* was actually genitival. It may mean simply that you could associate the name of *Werburh,* even after her death, in this loose fashion with the farm which she once possessed. The farm which once belonged to *Werburh* may have been more vaguely associated with her name by calling it *Werburgingwic* after her death. To this extent I find myself in agreement with the criticisms passed by Bradley on the strictly possessive theory in a review of the *Place-names of Northumberland*[1], though I can see no authority for his going on to state that *Werburgingwic* might mean "Werburh's people's dwelling," for at an earlier stage in the article he himself had pointed out, and quite rightly, that the difference between an OE name of the type *Aegel-*

[1] EHR 36, 294.

byrhtingahyrst and of the type *Alfredingtun* is the same as that between "the Johnsons' estate" and "the Johnson estate." To me this last comparison puts the case perfectly. The link between *Johnson* and *estate* in the latter phrase is exactly the link which I conceive to exist between *Werburh* and *wic* and between *Heregraue* and *feld.* It may be added that Karlström also takes the view that the distinction between the *ingtun* and *ingatun* type is fundamental[1].

When in the Introduction already referred to I laid stress on the importance of distinguishing *ing* and *inga* in Old English names, I tried to emphasise the point by arguing that it was not likely that one would have a place called *Cyneburgingtun* if it meant "farm of Cyneburh's people," this form being for earlier *Cyneburgingatun*, a patronymic formation from a woman's name. This was I think a mistake, and that for two reasons. One had not realised the large number of place-names containing women's names which bear witness to the prominent part which women might play in Old English society, and it laid too great stress on the strict interpretation of *inga* as a patronymic. We are not committed to believing in matriarchy among the Anglo-Saxons by believing in the possibility of such a name as *Cyneburgingatun*, for the *inga*(*s*)-suffix need mean nothing more than "the people that have to do with," not necessarily the "children of." Recently a very curious case has come to light in Sussex where we seem to have clear evidence of

[1] *Op. cit.* 20.

an *inga*-formation in connexion with a woman's name. For *Hardham* near Fittleworth we have two distinct series of forms, one of the very few cases in which one can really speak of two "types" for a place-name. From the time of Domesday to 1380 we have a series of forms *Heriedeham, Herham, Herietham, Herietheham, Erytham.* From the reign of Stephen down to 1724 we have a series *Eringeham, Heringham, Herrynggeham, Heryngham, Heryngeham, Herryngham* with such forms as *Herryngham al. dicta Herdham* (1399), *Hardham al. Heringham* (1602), *Hardham* otherwise *Irringham* (1740) which show the complete equivalence of the series. The first series can only be from OE *Heregȳðehām,* "homestead of a woman named *Heregȳð*." The second must be either from OE *Heringaham,* "homestead of Here's people," *Here* being a shortened or pet-form of *Heregȳð*, or, just conceivably, *Eringeham* may be a reduction of earlier *Heregȳðingaham,* though I do not think this is very likely. The name is of interest, first as giving us a name which in one of its forms at least shows us an *inga*(*s*)-formation from a woman's name; secondly, and more important, as showing that you could have two forms of a place-name, one embodying the name of the holder of the estate at some particular time and the other embodying the name of the whole group who gathered around the holder. One would very much like to know whether there is really any distinction in age between the two names or whether they came into use much about the same time.

We may now perhaps return to a further point in connexion with the names that have *ing* as distinct from *inga* as the connecting element. Bradley once held the view, which he afterwards withdrew in face of the evidence, that the choice between *ingtun* and *ingatun*, between *ingham* and *ingaham*, depended upon the length of the name to which it was added. This view, viz. that the choice between *ingtun* and *ingatun* was a matter of syllables rather than significance, has been revived lately, largely under the influence of Borowski's thesis on *Lautdubletten* in Old English[1]. One aspect of his theory is that a medial syllable may at times disappear so as to keep the number of syllables in the various forms of a noun-paradigm at the same figure. Thus you will tend to have OE *Sætęrnesdæg* (nominative) but *Sæterndæge* (dative). Ekwall in a recent paper[2], making a notable contribution to our knowledge of the inflexional forms of OE place-name elements, showed that all the evidence of good early charters and other reliable documents was in favour of believing that the true locatives of *ham* and *wic* in OE were still preserved in the uninflected forms *ham* and *wic* found in place-name forms as against the dative-

[1] The place-name forms *Beardanea*, *Beardsætan*, with which Borowski makes much play, were an unfortunate choice, for it is well known to place-name students that a name of the type *Beardsætan*, "settlers belonging to Bardney," was formed from *Beardanea* by the process of adding *sæte* direct to the stem. It does not stand for earlier *Beardansætan*.

[2] *Namn och Bygd*, 16, 59 ff.

locative *tune* found under like conditions in connexion with *tun*. He went on to suggest that herein might lie the secret of the admitted fact that *ingatun* forms are as rare in connexion with *tun* as *ingham* forms are rare in connexion with *ham*. The *Lytlingatun*-forms of the nominative would disappear in favour of the more commonly occurring dative forms *Lytlingtune* and ultimately the latter would prevail. On the other hand, in *Lytlingaham*, as there was normally no suffix, either in the nominative or in the locative, the *inga* would remain. This idea that *ing* must normally go back to earlier *inga* raised such important questions and seemed so definitely in contradiction to what one had recently observed with regard to the pure connective *ing* in Sussex, where in some cases as we have seen it could scarcely have been for earlier *inga*, that it seemed worth while to examine the OE evidence as a whole on this point. The task, with the aid of Karlström's volume, is not a very difficult one.

Taking in the first instance all the examples of *ingtun*- and *ingatun*-names found in would-be OE documents (excluding names of doubtful etymology and a good number of which the forms are obviously ME), one finds 120 or more which show *ingtun* and nothing else, and some dozen which show signs of being *ingatun*-formations. Of the *ingtun*-names some 29 are recorded only in the nominative, some 26 in both nominative and dative, and some 54 in the dative only. The rest are doubtful, having either *tune* in the nominative or *tun* in the dative. This might be taken merely as a mark of the thoroughness of the levelling-out process, but detailed examination of the forms of the names which show signs of *ingatun* reveals that this is not the case. Practically all the evidence tends in the opposite direction. For Teddington (Mx) we have nom. *Tudingtun, Tudintun* but dat. *Tudincgatunæ.* Washington (Sx) is only found in the dat. and always has medial *a* as in *Wessingatunæ.* Catherington (Ha) has dat. forms *Cat(e)ringatune* and *Cateringatun* indifferently, and Hoddington (Ha) has *Hoddingatun* (nom.), *Hoddingatune* (dat.). Watlington (O) has dat. *Wæclinctune* and *Hwætlingatune,* but in the nominative *Wæclingtun.* For Quarrington (L) the one OE form is *æt Coringatune* (dat.) and for Winterton (L) is *æt Wintringatune.* For Lotherton (Y) we have nom. *Luteringtun, Luttringtun* but dative *Luteringatune,* and for Markington (Y) we have nom. *Mercingatun,* dat. *Mercingatune.* Knedlington (Y) and Eastrington (Y), for each of which we have only one form, so far as they go, do conform to the suggested principle, with forms *Cnyllingatun* and *Eastringatun* in the nominative. Willington (Du) has indifferently *Twilingatun, Twinlingtun.*

These forms are derived from documents of varying age and reliability, but for a good number of the names we have forms derived from originals:—2 of the 8th century, 7 of the 9th, 18 of the 10th, with 4 more from the Pershore charter which, though copied in the 11th, probably retains the forms of the 10th century. It cannot be a mere chance that in all these names (drawn from whatever source) the evidence in every case except two tends to contradict the theory, and that in some cases it is directly opposed to it.

Of the various other second elements the only ones which show a clear preference for *ing* as against *inga* are *land, denn* and *wic*. We have some 30 or more examples of *land*-compounds and only one, the county-name *Westmoringaland,* shows *inga.* Of these *land*-names some 23 are found in the nominative, four are found both in the nominative and dative, and three others in either the dative or the genitive. For a large proportion of the names we have forms derived from originals:—3 of the 8th century, 14 of the first half of the 9th century and 2 of the 10th century.

For *denn* we have some 29 *ing*-compounds as against a possible one or two with *inga.* Of these 25 are nominative forms, 1 is a dative, 2 are both nominative and dative. Three are derived from originals of the 8th century, 14 from originals of the first half of the 9th century and 4 from originals of the second half of that century, 5 from originals of the 10th century. *land* and *denn* are imparisyllabic as between nominative and dative and yet we have no sign of any variation between *ing* and *inga* in these cases.

wic always has *ing* and not *inga,* and as it had a locative *wic* it should of course show persistent *inga* like the *ingaham* names, but, as Ekwall points out, the inflected dative plural was often used instead of the locative singular, so that one might argue that the dative plural forms led to the extinction of all *inga*-forms and its evidence against the theory must not be pressed.

Other second elements, whether parisyllabic like *burna* or imparisyllabic like *leah,* show no marked preference for formations of one type or the other, though it may be noted that *ing*-formations are much more numerous than the *inga*-ones. Several names might be quoted which in their forms definitely contradict the alleged principle.

How are we to account then for the marked preference of *ham* for *inga*-formations and *tun*, *land* and *denn* for *ing*-ones if it is not a matter of form? The main answer is I think provided by Ekwall's own explanation of the survival of the locative *ham* as against the dative-locative *tune,* viz. that the *ham*-names are of an essentially older type than the *tun*-names. That view is of course supported by the distribution of the *ingham*-names as revealed in his *Place-names in -ing* and by the distribution of *ham* and *tun* throughout the country. The farther we move away from the areas of earliest settlement, the fewer become the *hams* and the more numerous the *tuns.*

There is also a further point. There is no doubt that Ekwall was right in postulating an early age for the *ingas*- and *ingaham*-names in his *Place-names in -ing.* It is reasonable to believe that as the settlement advanced the old tribal, clan, or folk-system, or whatever we like to call it, implied in the *ingas*-, *inga*-names, was gradually fading away and the individual loomed more prominently. By the time that *tun* had become the great habitative suffix there were probably few such groups left, and the preference for *ingtun* as against *ingatun* records the new social conditions.

One other factor has however to be taken into consideration. Certain places would naturally be associated with a whole community, others with an individual person. The very early use in Kent of *land* and *denn* always with *ing* rather than *inga* must probably be explained in this way. *Inga*-formations were already in common use. If you could not speak of an *inga-denn* or an *inga-land*, the reason would seem to lie in some difference in the social conditions under which the *denn* or the *land* were occupied as against those prevailing for the *ham*.

We have just been speaking of group-names. It is perhaps of interest in conclusion to note one or two pieces of evidence which have recently come to light in Sussex and have definite bearing, as it seems to me, on their use. Ekwall noted that *Palinga schittas*, a swine-pasture of Felpham, now known to be in the neighbourhood of Petworth, might naturally be associated with the *Palingas* of Poling, some five miles to the east of Felpham. So also *Goringlee* in West Chiltington may have been the woodland of *Goring*, both containing OE *Garingas*[1]. The parishes of

[1] The possibility of this would be very much increased, indeed made certain, if we could be sure that Karlström is right in identi-

Upper *Beeding* and Lower *Beeding*, the former by the Downs, the latter up in the Weald, are separated by some miles. They contain an *ingas*-name and in former times were one parish. Lower *Beeding* is the Wealden settlement made from Upper *Beeding*. The most interesting piece of evidence however is less obvious. In the boundaries of Washington we have mention of a place called *horningacumb* of which it is difficult to determine the exact site. Among the swine-pastures of Washington was included Horsham. In Horsham there is a small place called *Horn Brook* of which the early forms are *Horningebrok*, etc. Names in *horn* or *Horn* are so rare in Old English nomenclature that it seems almost certain that the *Horningas* who gave their name to the combe in Washington also had their piece of *brook* or water-meadow in Horsham and gave their name to it. All these names must have arisen while the *ingas*-group was still a living reality. This piece of evidence has its interest in its bearing on another problem. It all fits in with the other place-name evidence which suggests that the settlement of the Wealden area took place at a quite early date. As Hilaire Belloc has reminded us more than once, the Weald was not the impenetrable mass of tangled thicket and forest that it has sometimes been thought, and the South Saxons soon made their way right into the heart of it.

fying the *Derantun* of BCS 702 with Durrington (Sx), *v. supra* 23. This place had a swine-pasture called *garungaleah* which is identical, at least in form, with *Goringlee*, and *Goring* adjoins Durrington.

And so the circle is completed. We began with the problems of the settlement and we come back to them in the end. As I have written these lectures I have realised more fully perhaps than ever before how difficult is the path of the place-name student. Progress is slowly being made but there is much material still to be gathered, there are many problems which still remain unsolved. Much is being done by scholars both at home and abroad in helping us to overcome these difficulties, but we must realise that there are many problems, not only of the individual name, but of whole classes of place-name phenomena, of which the ultimate secret may elude us for lack of the requisite data for solving them. We have Old English material of anything more than negligible quantity for little more than a third of our English counties. That material makes it abundantly clear, first, that speculation based upon post-Conquest material alone may at times go sadly astray, and, secondly, that names do not always become easier even when we can carry them back to their Old English forms. We are bound therefore again and again to be hazarding solutions about which there can be no certainty or finality. There is probably very little Old English material still to be brought to light, but there is abundance of good unpublished Middle English material and each new batch, as oft-repeated experience shows, may contain something which must make us re-shape our views upon this or that problem. To show that this is so has been one of the chief aims of these lectures, and as I think over them I am sure that, if we were ruthlessly self-critical and logical, these lectures and our place-name books generally would have to be sprinkled with the saving words, "possibly," "probably," "perhaps," far more even than they are at present, and certainly far more than the reading or listening public would ever tolerate. We want to guard specially against the danger of thinking that because a thing "may be" therefore it "must be" in the realm of place-name study. If we fall into that error there will inevitably be lying in wait for us that missing piece of comparative or topographical evidence, that undiscovered form, that long arm of coincidence which will be our ultimate undoing.

ADDENDUM TO LECTURE III

In explanation of the personal name *Grante* (p. 103), it has been suggested recently (ZONF 5, 187) that this is the French nickname *le Grant, le Grand,* but this cannot hold, for the name in the Croyland Cartulary is a single name and not the second of two names and, it may be added, even if it were a second name, would not appear without the definite article in a twelfth-century name.

INDEXES

(a) *Index rerum*

(b) *Index locorum*

(c) *Index nominum*

Names not found on independent record are marked with a single star if their existence can be safely inferred. Such names may be regarded as hardly less certain than the unstarred ones. Those inferred on less certain grounds are marked with a double star.

(i) OE

(ii) Scandinavian

(*d*) *Index verborum*

Words marked with a star are not on independent record.

(i) OE

(ii) ON and ODan

(iii) Modern English (chiefly dialectal

www.ingramcontent.com/pod-product-compliance
Ingram Content Group UK Ltd.
Pitfield, Milton Keynes, MK11 3LW, UK
UKHW040656180125
453697UK00010B/207